Motorbooks International

CHEVROLET
PICKUP
RED BOOK

Peter C. Sessler

First published in 1993 by Motorbooks International Publishers &
Wholesalers, PO Box 2, 729 Prospect Avenue, Osceola, WI 54020 USA

The information in this book is true and complete to the best of our
knowledge. All recommendations are made without any guarantee on
the part of the author or Publisher, who also disclaim any liability
incurred in connection with the use of this data or specific details

We recognize that some words, model names and designations, for
example, mentioned herein are the property of the trademark holder.
We use them for identification purposes only. This is not an official
publication

Motorbooks International books are also available at discounts in
bulk quantity for industrial or sales-promotional use. For details
write to Special Sales Manager at the Publisher's address

Library of Congress Cataloging-in-Publication Data
Sessler, Peter C.
    Chevrolet pickup red book / Peter C. Sessler.
      p. cm.—(Motorbooks International red book series)
    Includes index.
    ISBN 0-87938-771-8
    1. Chevrolet trucks—History. 2. Chevrolet trucks—
Specifications. 3. Chevrolet trucks—Collectors and collecting.
I. Title. II. Series.
TL230.5.C45S47   1993
629.223—dc20   93-25847

On the front cover: The 1957 Chevrolet 3100 pickup truck
owned by Bob and Debbie Higgins of Davie, Florida. *Mike
Mueller*

On the back cover: The 1953 Chevrolet pickup. *Paul
McLaughlin*

Printed and bound in the United States of America

Contents

Special thanks to Joe Bertalan, Ed Conrads, Max and Mike McGuire, and Chevrolet Motor Division

Introduction

The *Chevrolet Pickup Red Book* is designed to help the Chevy enthusiast determine the authenticity and originality of any Chevrolet light truck built between 1946 and 1979. 1992. Also included are the commercial passenger cars produced during this period, such as the sedan delivery, El Camino, Corvair, Vega Panel Express, and Chevy LUV minipickup. Each chapter covers a model year. Included are production figures, VIN serial number decoding information, engine suffix codes, rear axle codes, transmission codes, exterior colors, pricing, options, and selected facts. Because of space limitations, not all the information is included for all trucks, such as options and colors for the Corvair, Vega, and LUV. Coverage is not quite as limited on the El Camino, for which much of the same information has already been printed in the *Chevrolet SS Muscle Car Red Book*.

Chevrolet trucks from whatever period you're looking at are unquestionably popular. Total annual sales were always in the hundreds of thousands. This book covers trucks built during the last year of the art deco period,1946, followed by the Advance Design Series trucks built from 1947—55. The Task Force era began with the redesigned trucks introduced in 1955 and continued to 1959. The V-8 era of trucks began in 1960 and ended in 1966, and was followed by the Custom Sports Truck era of 1967—72. Modern-era trucks were those produced from 1973 to 1987. In each period, you'll find that once the basic body-and-chassis platform was introduced, it continued until the end of the period. Unlike cars, pickup trucks displayed few year-to-year styling and drivetrain changes. This started to change in the 1970s, when new segments of the buying public "discovered" trucks. Trucks became more carlike in terms of creature comforts and options and yet still retained their utilitarian usefulness qualities. Because the truck platforms did not lend themselves to styling changes every two or three years, Chevrolet resorted to different paint schemes, exterior graphics, and slight detail changes to maintain the image of "newness" each year.

Not all Chevrolet trucks were popular. The innovative Corvair pickups of the early 1960s are an example. The Vega-based Panel Express, although a good-looking truck, was too small, and Vega's engine problems didn't help either. The El Camino, however, proved to be a successful hybrid of a car and a truck.

This book covers only the 1/2-, 3/4-, and 1-ton light-duty trucks produced by Chevrolet. Not included are vans and any of the heavy-duty and over-1-ton models. Also not covered are the countless aftermarket conversions. Chevrolet sold to outside companies platforms that included the cab (or a variation thereof), chassis, and drivetrain. These companies would then install their own bodies, such as ambulances, campers, and many others.

For the enthusiast, the most important number in any Chevrolet truck is its vehicle identification number (VIN). The number has been changed several times since 1946, each time denoting a little more information. The changes are documented in the Appendix. As Also important are the engine serial identification numbers and engine casting date codes, which are stamped on the engine itself. Every Applicable engine code is codes are listed in each chapter, and again, the Appendix explains what the numbers mean. Transmission and rear axle codes are also included in each chapter so that the rest of the truck's drivetrain can be authenticated.

Even if all the numbers match on a particular car you are looking at, it will be to your advantage if the truck is documented. It is all the better if the previous owner can provide you with the original invoice or window sticker, any service records, the Protect-O-Plate, and the like.

The colors listed in each chapter are correct as far as they go. However, Chevrolet did build cars in colors and trim combinations not listed. As with all the information given here, be open to the possibility that exceptions can and do occur. This means you'll have to work harder hard to determine authenticity.

Every effort has been made to make sure the information contained in this book is correct. However, I would like to hear from any enthusiast with corrections or interesting additions. Please write to me care of Motorbooks International.

Chapter 1

1946 Trucks

Production

Calendar year	Total
1946	270,140
1947	335,343*

*includes late 1946s

Serial numbers

Description

3DPA00001

3—Assembly Plant code: 1—Flint MI, 2—Tarrytown NY, 3—St. Louis MO, 5—Kansas City MO, 6—Oakland CA, 8—Atlanta GA, 9—Norwood OH, 14—Baltimore MD, 20—Van Nuys CA, 21—Janesville WI

DP—Series: DJ—Passenger (Stylemaster) 1500 Series, CK & DP (Second Design) Commercial ½ ton 3100 Series, DR—Commercial ¾ ton 3600 Series DS—Commercial 1 ton 3800 Series

A—Month of production, A—January, B—February, C—March, D—April, E—May, F—June, G—July, H—August, I—September, J—October, K—November, L—December

00001—Consecutive Sequence Number

Location: On an underhood plate located on the right hand side of the cowl

Model, Wheelbase & GVW
1500 Series

Model Number and Description	Wheelbase (in)	GVW
1508 Sedan delivery (Stylemaster)	116	4,100
1514 Coupe pickup	116	4,100

3100 Series

	Wheelbase (in)	GVW
3102 Chassis w/flat face cowl	115	4,600
3103 Chassis and cab	115	4,600
3104 Pickup	115	4,600
3105 Panel	115	4,600
3106 Carryall suburban w/panel rear doors	115	4,600
3107 Canopy express	115	4,600
3112 Chassis w/cowl and windshield	115	4,600
3116 Carryall suburban w/end-gate	115	4,600
3122 Stripped chassis w/flat-face cowl	115	4,600
3132 Stripped chassis w/cowl and windshield	115	4,600

3600 Series

3602	Chassis w/flat-face cowl	125¼	5,800
3603	Chassis and cab	125¼	5,800
3604	Pickup	125¼	5,800
3605	Panel	125¼	5,800
3608	Chassis and cab w/platform body	125¼	5,800
3809	Chassis and cab w/stake body	125¼	5,800
3612	Chassis w/cowl and windshield	125¼	5,800
3622	Stripped chassis w/flat-face cowl	125¼	5,800
3632	Stripped chassis w/cowl & windshield	125¼	5,800

3800 Series

3800	Chassis extra length station wagon	134½	8,800
3802	Chassis w/flat-face cowl	134½	8,800
3803	Chassis and cab	134½	8,800
3804	Pickup	134½	6,700
3805	Panel	134½	6,700
3807	Canopy express	134½	6,700
3808	Chassis and cab w/Platform body	134½	8,800
3812	Chassis w/cowl and windshield	134½	8,800
3822	Stripped chassis w/flat-face cowl	134½	8,800
3832	Stripped chassis w/cowl & windshield	134½	8,800

Engine & Transmission Suffix Codes

Sedan Delivery
DAA—216.5ci I-6 1bbl 90hp—three-speed manual
DAM—216.5ci I-6 1bbl 90hp—three-speed manual

Truck—¹/₂ ton & Interim CK Series
BD—216.5ci I-6 1bbl 90hp—three-speed manual

¹/₂ ton—3100 Series, ³/₄ ton 3600 Series, 1-ton 3800 Series
DBA—216.5ci I-6 1bbl 90hp—three-speed manual
DBM—216.5ci I-6 1bbl 90hp—three-speed manual

Transmission Codes

Code	Type	Plant
1500 Series		
DA	3spd	Saginaw
DB	3spd	Muncie
DC	3spd	Toledo
DD	3spd	Saginaw
DE	3spd	Muncie
DF	3spd	Toledo
DG	3spd/HD	Saginaw
DH	3spd/HD	Muncie
DJ	3spd/HD	Toledo
DK	3spd/HD	Saginaw
DL	3spd/HD	Muncie
DM	3spd/HD	Toledo

3100 Series		
DN	3spd	Saginaw
DO	3spd	Muncie
DP	3spd	Toledo
3600 Series		
DQ	3spd	Saginaw
DR	3spd	Muncie
DS	3spd	Toledo
3800 Series		
PA	4spd	Saginaw
PB	4spd	Muncie
PC	4spd	Toledo
3100, 3600 Series		
PK	4spd	Saginaw

| PL | 4spd | Muncie |
| PM | 4spd | Saginaw |

Axle Identification

Code	Ratio	Plant

1500 Series
DA	4.11:1	Detroit
DB	4.11:1	Tonawanda
DC	3.73:1	Detroit
DD	3.73:1	Tonawanda

3100 Series
| DE | 4.11:1 | Detroit |
| DF | 4.11:1 | Tonawanda |

3600 Series
| DG | 4.57:1 | Detroit |
| DH | 4.57:1 | Tonawanda |

3800 Series
PA	5.14:1 (dual wheels)	Detroit
PB	5.14:1 (dual wheels)	Tonawanda
PC	5.14:1 (single wheel)	Detroit
PD	5.14:1 (single wheel)	Tonawanda

Exterior Color Codes
Apple Green	234C
Armour Yellow	234B
Omaha Yellow	234F
Hollywood Tan	234G
Airedale Brown/Circassian Brown	234H*
Bordeaux Maroon	234J
White	234L
Brewster Green	Std.
Boatswain Blue	234M
Export Blue	234E
Swift Red	234A
Cream Medium	234D
Black	234K

*Carryall Suburban only

Regular Production Options
1508 Sedan Delivery	$1,173

3100 Series
3102 Chassis w/f.f.cowl	644(796)

3103 Chassis and cab	679(922)
3104 Pickup	784(963)
3105 Panel	869(1,077)
3106 Carryall suburban w/panel rear doors	1,014(1,283)
3107 Canopy express	904(1,126)
3112 Chassis w/cowl and windshield	754(817)
3116 Carryall suburban w/end-gate	1,104(1,281)

Parentheses indicates 1946 Second series prices

3600 Series
3602 Chassis w/flat-face cowl	891
3603 Chassis and cab	1,016
3604 Pickup	1,069
3605 Panel	1,212
3608 Chassis and cab w/platform body	1,101
3809 Chassis and cab w/stake body	1,127
3612 Chassis w/cowl and windshield	911

3800 Series
3803 Chassis and cab	892
3805 Panel	1,264
3807 Canopy express	1,318

Option Number
200	Double-acting shock absorbers
201	Economy rear axle
205	Low rear axle ratio
207	Long running boards and rear fenders
210	Rear-view mirrors and brackets
213	Panel-type rear bumper bar
215	Wheel and Tire carrier
216	Oil-bath air cleaner
224	Economy Engine
228	Screen equipment for canopy express trucks
227	Heavy-duty clutch
232	Frame Extensions
237	Oil filter
241	Governor
244	Propeller shaft guard
245	Bumper guards
247	Radiator grille guard
249	Tail and stop lamp equipment
250	Rear fender irons
256	16-quart cooling system
261	Sun shades (visors)

Facts

The first series of 1946 Chevrolet pickup production began on September 1, 1945, and continued through September 1, 1946. The vehicles produced during this period are known as "interim" models and are indicated as such by the designation CK in the serial number. Second-series 1946 pickups were built from May 1, 1946, to May 3, 1947, and had a DP designation. You'll find some interim models titled 1945s and some second-series titled 1947s.

The 1946 pickups were basically a continuation of the trucks introduced in 1941. They were distinguished by a wood-floored cargo box with steel skid strips, black fenders, and full running boards. The bumpers were chrome.

All were powered by the 216.5ci inline six-cylinder engine rated at 90hp. Compression was 6.5:1. The standard transmission was a three-speed manual, and a four-speed manual was optional.

The 1/2-ton models used a semifloating rear axle, and all others used a full-floating design.

Although similar to the 1944—45 trucks, the 1946s benefited from six-ply tires, added fender reinforcing, and a better grade of vinyl for the seat covering. Leather was seats were optional.

Sedan and panel delivery models were based on the regular Chevrolet passenger cars. They were two-door models vehicles with a rear curbside opening door that opened toward the curb. Standard equipment was a single driver's bucket seat, and an auxiliary passenger's bucket seat was optional. They These models were powered by the same 216.5ci six-cylinder engine used in all 1946 models.

The Suburban Carryall used the same basic body as the panel truck but with rear side windows. It had two front doors, a tailgate hinged at the bottom, hingled tailgate and a liftgate hinged at the. top hinged liftgate. Panel doors would be available in 1950.

Chapter 2

1947 Trucks

Production

Calendar year	Total
1947	335,343*

*includes late 1946s

Serial numbers

Description

3EJA00001

3—Assembly Plant Code: 1—Flint MI, 2—Tarrytown NY, 3—St. Louis MO,
   5—Kansas City MO, 6—Oakland CA, 8—Atlanta GA, 9—Norwood OH,
   14—Baltimore MD, 20—Van Nuys CA, 21—Janesville WI

E—Year, E—1947

J—Series: J—1500, P—3100, R—3600, S—3800

A—Month of production, A—January, B—February, C—March, D—April,
   E—May, F—June, G—July, H—August, I—September, J—October,
   K—November, L—December

00001—Consecutive Sequence Number

Location: On a plate located on the left door lock pillar. Sedan Delivery serial number is under the floor mat on the right side.

Model, Wheelbase & GVW

Model Number and Description	Wheelbase (in)	GVW
1500 Series		
1508 Sedan delivery	116	4,100
3100 Series		
3102 Chassis w/flat face cowl	115	4,600
3103 Chassis and cab	115	4,600
3104 Pickup	115	4,600
3105 Panel	115	4,600
3106 Carryall suburban w/panel rear doors	115	4,600
3107 Canopy express	115	4,600
3112 Chassis w/cowl and windshield	115	4,600
3116 Carryall suburban w/end-gate	115	4,600
3122 Stripped chassis w/flat-face cowl	115	4,600
3132 Stripped chassis w/cowl and windshield	115	4,600

3600 Series

3602	Chassis w/flat-face cowl	125$\frac{1}{4}$	5,800
3603	Chassis and cab	125$\frac{1}{4}$	5,800
3604	Pickup	125$\frac{1}{4}$	5,800
3605	Panel	125$\frac{1}{4}$	5,800
3608	Chassis and cab w/platform body	125$\frac{1}{4}$	5,800
3809	Chassis and cab w/stake body	125$\frac{1}{4}$	5,800
3612	Chassis w/cowl and windshield	125$\frac{1}{4}$	5,800
3622	Stripped chassis w/flat-face cowl	125$\frac{1}{4}$	5,800
3632	Stripped chassis w/cowl & windshield	125$\frac{1}{4}$	5,800

3800 Series

3802	Chassis w/flat-face cowl	134$\frac{1}{2}$	8,800
3803	Chassis and cab	134$\frac{1}{2}$	8,800
3804	Pickup	134$\frac{1}{2}$	6,700
3805	Panel	134$\frac{1}{2}$	6,700
3807	Canopy express	134$\frac{1}{2}$	6,700
3808	Chassis and cab w/Platform body	134$\frac{1}{2}$	8,800
3812	Chassis w/cowl and windshield	134$\frac{1}{2}$	8,800
3822	Stripped chassis w/flat-face cowl	134$\frac{1}{2}$	8,800
3832	Stripped chassis w/cowl & windshield	134$\frac{1}{2}$	8,800

Engine & Transmission Suffix Codes

1500 Series

EAA—216.5ci I-6 1bbl 90hp—three-speed manual
EAM—216.5ci I-6 1bbl 90hp—three-speed manual
EAC—216.5ci I-6 1bbl 90hp—three-speed manual HD
EAP—216.5ci I-6 1bbl 90hp—three-speed manual HD

3100 Series

EBA—216.5ci I-6 1bbl 90hp—three/four speed manual
EBM—216.5ci I-6 1bbl 90hp—three/four speed manual
BECA—216.5ci I-6 1bbl 90hp—three/four speed manual HD
BECM—216.5ci I-6 1bbl 90hp—three/four speed manual HD
AECA—216.5ci I-6 1bbl 90hp—three/four speed

3600 Series

AECM—216.5ci I-6 1bbl 90hp—three/four speed

3800 Series

AECM—216.5ci I-6 1bbl 90hp—three/four speed

Transmission Codes

Code	Type	Plant
1500 Series		
EA	3spd	Saginaw
EB	3spd	Muncie
EC	3spd	Toledo
ED	3spd	Saginaw
EE	3spd	Muncie
EF	3spd	Toledo
EG	3spd/HD	Saginaw
EH	3spd/HD	Muncie
EJ	3spd/HD	Toledo
EK	3spd/HD	Saginaw
EL	3spd/HD	Muncie
EM	3spd/HD	Toledo
3100 Series		
EN	3spd	Saginaw
EO	3spd	Muncie
EP	3spd	Toledo

3600 Series

EQ	3spd	Saginaw
ER	3spd	Muncie
ES	3spd	Toledo

3800 Series

QA	4spd	Saginaw
QB	4spd	Muncie
QC	4spd	Toledo

3100, 3600 Series

QK	4spd	Saginaw
QL	4spd	Muncie
QM	4spd	Saginaw

Axle Identification

Code	Ratio	Plant

1500 Series

EA	4.11:1	Detroit
EB	4.11:1	Tonawanda
EC	3.73:1	Detroit
ED	3.73:1	Tonawanda

3100 Series

| EE | 4.11:1 | Detroit |
| EF | 4.11:1 | Tonawanda |

3600 Series

| EG | 4.57:1 | Detroit |
| EH | 4.57:1 | Tonawanda |

3800 Series

QA	5.14:1 (dual wheels)	Detroit
QB	5.14:1 (dual wheels)	Tonawanda
QC	5.14:1 (single wheel)	Detroit
QD	5.14:1 (single wheel)	Tonawanda

Exterior Color Codes

Forester Green	STD*
Swift Red	234A
Armour Yellow	234B
White	234C
Jet Black	234D
Omaha Orange	234E
Cape Maroon	234F
Mariner Blue	234G
Windsor Blue	234H
Seacrest Green	234J
Sun Beige	234K
Cream Medium	234L
Maryland Black	STD**
Oxford Maroon	235A
Lullwater Green	235B
Battleship Grey	235C
Sport Beige	235D
Freedom Blue	235E
Lullwater Green/Sport Beige	235F
Lullwater Green/Lakeside Green	235G
Freedom Blue/Ozone Blue	235H
Scout Brown/Sport Beige	235J
Fathom Green/Channel Green	STD***

*all except Carryall Suburban
**Sedan Delivery
***Carryall Suburban only

Regular Production Options

| 1508 Sedan Delivery | $1,233 |

3100 Series

3102 Chassis w/ flat-face cowl	843
3103 Chassis and cab	1,030
3104 Pickup	1,087
3105 Panel	1,238
3107 Canopy express	1,289
3112 Chassis w/cowl and windshield	863
3116 Carryall suburban w/end-gate	1,474

3600 Series

3602 Chassis w/flat-face cowl	941
3603 Chassis and cab	1,128
3604 Pickup	1,201
3608 Chassis and cab w/platform body	1,211
3809 Chassis and cab w/stake body	1,258
3612 Chassis w/cowl and windshield	962

3800 Series

3802 Chassis w/flat-face cowl	988
3803 Chassis and cab	1,176
3804 Pickup	1,279
3805 Panel	1,445
3807 Canopy express	1,523
3808 Chassis and cab w/Platform body	1,295
3809 Chassis and cab w/stake body	1,362

Option Number

200 Double-acting shock absorbers
201 Economy rear axle
210 Rear-view mirrors and brackets, LH/RH
216 Oil-bath air cleaner
227 Heavy-duty clutch
232 Frame extensions
237 Oil filter
241 Governor
245 Bumper guards
249 Tail and stop lamp equipment
256 17½ quart cooling system (HD radiator)
263 Right-hand front seat
268 Double-acting rear springs
271 Radiator overflow return tank
316 HD 3-speed transmission
318 HD 4-speed transmission
348 Tru-stop brake equipment
379 Inside fuel tank
385 Fresh-air ventilator, heater & defroster
386 Chrome radiator grille
387 Rear-corner windows
389 Wide running boards
390 Deluxe cab & equipment

Facts

The 1947 Chevrolet truck line was divided into two series: Thriftmaster and Loadmaster. The Thriftmasters comprised trucks up to 1 ton, and the Loadmasters were 1-1/2 tons and larger. Both shared the same styling, although the Loadmasters were larger. As a group, 1947—55 Chevrolet trucks are known as the Advance Design Series trucks.

For 1947, the Chevrolet pickups were redesigned and updated: they were longer, lower, and wider. The cab was 8in wider and 7in longer. The hood was hinged at the cowl, enabling it to be opened from the front, replacing the previous side-opening version. The front grille was distinguished by five large horizontal bars that bulged in the center. Standard trucks had a painted grille, whereas Deluxe models got a chrome version. The same finish differentiation applied to the front bumper. The wheels were painted black and used a chrome hubcap stamped with a red Chevrolet name. Suburbans came with body-colored painted wheels.

All trucks used a Chevrolet emblem on each side of the hood near the cowl.

Pickup boxes were available in three sizes, measuring 78in, 87in, and 108in long inside for 1/2-, 3/4-, and 1-ton pickups. They all had the same 50in inside width. The floors were covered with longleaf yellow pine.

The standard color for all models was Forester Green, except for the Suburban, which came with a two-tone combination of Fathom Green and Channel Green. The interior was finished in silver gray with burgundy Naugahyde. Leather was optional. The rooflining headlining on all models was made of fiberboard, painted to match the interior color. On the panel and Canopy Express models, the covering extended only to the driver's compartment, whereas the Suburbans got a full-length rooflining headlining.

Optional were rear quarter windows. These enhanced rear vision and were also part of the Deluxe Cab option.

The standard and only engine was the Chevrolet 216.5ci inline six rated at 90hp and mated to a three-speed manual transmission.

These trucks were equipped with a tool kit that consisted of three open-end wrenches, a spark plug wrench, a 6in round-shank screwdriver, a 6in pliers, and a 10oz ball-peen hammer. The hammer , which was painted; . All all the other tools were plated. They The tools were all enclosed in an envelope-type bag. In addition, each truck was equipped with a jack and a wheel wrench.

Chapter 3

1948 Trucks

Production

Calendar year	Total
1948	389,690

Serial numbers

Description

3FJA00001

3—Assembly Plant Code: 1—Flint MI, 2—Tarrytown NY, 3—St. Louis MO,
   5—Kansas City MO, 6—Oakland CA, 8—Atlanta GA, 9—Norwood OH,
   14—Baltimore MD, 20—Van Nuys CA, 21—Janesville WI

F—Year, F—1948

J—Series: J—1500, P—3100, R—3600, S—3800

A—Month of production, A—January, B—February, C—March, D—April,
   E—May, F—June, G—July, H—August, I—September, J—October,
   K—November, L—December

00001—Consecutive Sequence Number

Model, Wheelbase & GVW

Model Number and Description	Wheelbase (in)	GVW
1500 Series		
1508 Sedan delivery	115	4,100
3100 Series		
3102 Chassis w/flat face cowl	116	4,600
3103 Chassis and cab	116	4,600
3104 Pickup	116	4,600
3105 Panel	116	4,600
3107 Canopy express	116	4,600
3112 Chassis w/cowl and windshield	116	4,600
3116 Carryall suburban w/end-gate	116	4,600
3122 Stripped chassis w/flat-face cowl	116	4,600
3132 Stripped chassis w/cowl and windshield	116	4,600
3600 Series		
3602 Chassis w/flat-face cowl	125¼	5,800
3603 Chassis and cab	125¼	5,800
3604 Pickup	125¼	5,800
3608 Chassis and cab w/platform body	125¼	5,800
3809 Chassis and cab w/stake body	125¼	5,800

3612	Chassis w/cowl and windshield	125¼	5,800
3622	Stripped chassis w/flat-face cowl	125¼	5,800
3632	Stripped chassis w/cowl and windshield	125¼	5,800

3800 Series

3802	Chassis w/flat-face cowl	137	8,800
3803	Chassis and cab	137	8,800
3804	Pickup	137	6,700
3805	Panel	137	6,700
3807	Canopy express	137	6,700
3808	Chassis and cab w/platform body	137	8,800
3809	Chassis and cab w/stake body	137	8,800
3812	Chassis w/cowl and windshield	137	8,800
3822	Stripped chassis w/flat-face cowl	137	8,800
3832	Stripped chassis w/cowl and windshield	137	8,800

Engine & Transmission Suffix Codes

1500 Series

FAA—216.5ci I-6 1bbl 90hp—three-speed manual
FAM—216.5ci I-6 1bbl 90hp—three-speed manual
FAC—216.5ci I-6 1bbl 90hp—three-speed manual
FAP—216.5ci I-6 1bbl 90hp—three-speed manual

3100 Series

FBA—216.5ci I-6 1bbl 90hp—three/four speed manual
FBM—216.5ci I-6 1bbl 90hp—three/four speed manual
BFCA—216.5ci I-6 1bbl 90hp—three/four speed manual
BFCM—216.5ci I-6 1bbl 90hp—three/four speed manual

3600 Series

AFCA—216.5ci I-6 1bbl 90hp—three/four speed manual
AFCM—216.5ci I-6 1bbl 90hp—three/four speed manual

3800 Series

AFCM—216.5ci I-6 1bbl 90hp—three/four speed manual

Transmission Codes

Code	Type	Plant
1500 Series		
FA	3spd	Saginaw
FB	3spd	Muncie
FC	3spd	Toledo
FD	3spd	Saginaw
FE	3spd	Muncie
FF	3spd	Toledo
FG	3spd/HD	Saginaw
FH	3spd/HD	Muncie
FJ	3spd/HD	Toledo
FK	3spd/HD	Saginaw
FL	3spd/HD	Muncie
FM	3spd/HD	Toledo

3100 Series		
FN	3spd	Saginaw
FO	3spd	Muncie
FP	3spd	Toledo
3600 Series		
FQ	3spd	Saginaw
FR	3spd	Muncie
FS	3spd	Toledo
3800 Series		
FA	4spd	Saginaw
FB	4spd	Muncie
FC	4spd	Toledo
3100,3600 Series		
RK	4spd	Saginaw

RL	4spd	Muncie	*all except Carryall Suburban
RM	4spd	Saginaw	**Sedan Delivery
			***Carryall Suburban only

Axle Identification

Code	Ratio	Plant

1500 Series
FA	4.11:1	Detroit
FB	4.11:1	Buffalo
FC	3.73:1	Detroit
FD	3.73:1	Buffalo

3100 Series
| FE | 4.11:1 | Detroit |
| FF | 4.11:1 | Detroit |

3600 Series
| FG | 4.57:1 | Detroit |
| FH | 4.57:1 | Buffalo |

3800 Series
RA	5.14:1 (dual wheels)	Detroit
RB	5.14:1 (dual wheels)	Buffalo
RC	5.14:1 (single wheel)	Detroit
RD	5.14:1 (single wheel)	Buffalo

Exterior Color Codes

Forester Green	STD*
Swift Red	234A
Armour Yellow	234B
White	234C
Jet Black	234D
Omaha Orange	234E
Cape Maroon	234F
Mariner Blue	234G
Windsor Blue	234H
Seacrest Green	234J
Sun Beige	234K
Cream Medium	234L
Maryland Black	STD**
Live Oak Green	235A
Lake Como Blue	235B
Dove Gray	235C
Silver Gray Green	235D
Battleship Grey	235E
Oxford Maroon	235F
Satin Green/Marsh Brown	235G
Dove Gray/Lake Como Blue	235H
Fathom Green/Channel Green	STD***

Regular Production Options

1508 Sedan Delivery	$1,361

3100 Series
3102 Chassis w/ flat-face cowl	890
3103 Chassis and cab	1,113
3104 Pickup	1,180
3105 Panel	1,377
3107 Canopy express	1,429
3112 Chassis w/cowl and windshield	910
3116 Carryall suburban w/end-gate	1,627

3600 Series
3602 Chassis w/flat-face cowl	1,004
3603 Chassis and cab	1,227
3604 Pickup	1,315
3608 Chassis and cab w/platform body	1,320
3809 Chassis and cab w/stake body	1,378
3612 Chassis w/cowl and windshield	1,025

3800 Series
3802 Chassis w/flat-face cowl	1,087
3803 Chassis and cab	1,310
3804 Pickup	1,425
3805 Panel	1,596
3807 Canopy express	1,674
3808 Chassis and cab w/platform body	1,440
3809 Chassis and cab w/stake body	1,513

Option Number

200 Double-acting shock absorbers
201 Economy rear axle
210 Rear-view mirrors and brackets, LH/RH
216 Oil-bath air cleaner
227 Heavy-duty clutch
237 Oil filter
241 Governor
245 Bumper guards
249 Tail and stop lamp equipment

256 17½ quart cooling system (HD radiator)
263 Right-hand front seat
268 Double-acting rear springs
271 Radiator overflow return tank
316 HD 3-speed transmission
318 HD 4-speed transmission
348 Tru-stop brake equipment

379 Inside fuel tank
385 Fresh-air ventilator, heater & defroster
386 Chrome radiator grille
387 Rear corner windows
389 Wide running boards
390 Deluxe cab & equipment

Facts

No exterior changes were made on the 1948 models. In the interior, the transmission shift lever was relocated to the steering column on three-speed-equipped 1/2- and 3/4-ton trucks. The parking brake was also relocated, to the floor.

The 216.5ci six-cylinder engine was modified through the use of thin-wall Babbitt main and rod bearings. These replaced the previous poured type. Other mechanical improvements included the use of a combination fuel-and-vacuum pump for better windshield wiper action.

The tool kit was no longer available, save for but the jack and wheel wrench were.

Another small modification was on the inside door handles. For 1948 and later models, the handle had to be pulled to open the door.

For safety reasons, treads were installed on the running boards.

1949 Trucks

Production

Calendar year	Total
1949	383,543

Serial numbers

Description

3GJA00001

3—Assembly Plant Code: 1—Flint MI, 2—Tarrytown NY, 3—St. Louis MO,
  5—Kansas City MO, 6—Oakland CA, 8—Atlanta GA, 9—Norwood OH,
  14—Baltimore MD, 20—Van Nuys CA, 21—Janesville WI

G—Year, G—1949

J—Series: J—1500, P—3100, R—3600, S—3800

A—Month of production, A—January, B—February, C—March, D—April,
  E—May, F—June, G—July, H—August, I—September, J—October,
  K—November, L—December

00001—Consecutive Sequence Number

Model, Wheelbase & GVW

Model Number and Description	Wheelbase (in)	GVW
1500 Series		
1508 Sedan delivery	115	4,100
3100 Series		
3102 Chassis w/flat-face cowl	116	4,600
3103 Chassis and cab	116	4,600
3104 Pickup	116	4,600
3105 Panel	116	4,600
3107 Canopy express	116	4,600
3112 Chassis w/cowl and windshield	116	4,600
3116 Carryall suburban w/end-gate	116	4,600
3122 Stripped chassis w/flat-face cowl	116	4,600
3132 Stripped chassis w/cowl and windshield	116	4,600
3600 Series		
3602 Chassis w/flat-face cowl	125¼	5,800
3603 Chassis and cab	125¼	5,800
3604 Pickup	125¼	5,800
3608 Chassis and cab w/platform body	125¼	5,800
3809 Chassis and cab w/stake body	125¼	5,800

3612	Chassis w/cowl and windshield	125¼	5,800
3622	Stripped chassis w/flat-face cowl	125¼	5,800
3632	Stripped chassis w/cowl and windshield	125¼	5,800

3800 Series

3802	Chassis w/flat-face cowl	137	8,800
3803	Chassis and cab	137	8,800
3804	Pickup	137	6,700
3805	Panel	137	6,700
3807	Canopy express	137	6,700
3808	Chassis and cab w/platform body	137	8,800
3809	Chassis and cab w/stake body	137	8,800
3812	Chassis w/cowl and windshield	137	8,800
3822	Stripped chassis w/flat-face cowl	137	8,800
3832	Stripped chassis w/cowl and windshield	137	8,800

Engine & Transmission Suffix Codes

1500 Series

GAA—216.5ci I-6 1bbl 90hp—three-speed manual
GAM—216.5ci I-6 1bbl 90hp—three-speed manual
GAC—216.5ci I-6 1bbl 90hp—three-speed manual
GAP—216.5ci I-6 1bbl 90hp—three-speed manual

3100 Series

GBA—216.5ci I-6 1bbl 90hp—three/four speed manual
GBM—216.5ci I-6 1bbl 90hp—three/four speed manual
BGCA—216.5ci I-6 1bbl 90hp—three/four speed manual
BGCM—216.5ci I-6 1bbl 90hp—three/four speed manual

3600/3800 Series

AGCA—216.5ci I-6 1bbl 90hp—three/four speed manual
AGCM—216.5ci I-6 1bbl 90hp—three/four speed manual

Transmission Codes

Code	Type	Plant

1500 Series

GA	3spd	Saginaw
GB	3spd	Muncie
GC	3spd	Toledo
GD	3spd	Saginaw
GE	3spd	Muncie
GF	3spd	Toledo
GG	3spd/HD	Saginaw
GH	3spd/HD	Muncie
GJ	3spd/HD	Toledo
GK	3spd/HD	Saginaw
GL	3spd/HD	Muncie
GM	3spd/HD	Toledo

3100 Series

GN	3spd	Saginaw
GO	3spd	Muncie
GP	3spd	Toledo

3600 Series

GQ	3spd	Saginaw
GR	3spd	Muncie
GS	3spd	Toledo

3800 Series

SA	4spd	Saginaw
SB	4spd	Muncie
SC	4spd	Toledo

3100,3600 Series

SK	4spd	Saginaw
SL	4spd	Muncie
SM	4spd	Saginaw

Axle Identification

Code	Ratio	Plant

1500 Series

GA	4.11:1	G&A
GB	4.11:1	Tonawanda

GC	3.73:1	G&A
GD	3.73:1	Tonawanda

3100 Series

GE	4.11:1	G&A
GF	4.11:1	Tonawanda

3600 Series

GG	4.57:1	G&A
GH	4.57:1	Tonawanda
GN	5.14:1	G&A
GP	5.14:1	Tonawanda

3800 Series

SA	5.14:1 (dual wheels)	
SB	5.14:1 (dual wheels)	G&A
SC	5.14:1 (single wheel)	Tonawanda
SD	5.14:1 (single wheel)	G&A
		Tonawanda

Exterior Color Codes

Forester Green	STD*
Maryland Black	STD**
Fathom Green/Channel Green	STD***
Swift Red	234A
Armour Yellow	234B
White	234C
Jet Black	234D
Omaha Orange	234E
Cape Maroon	234F
Mariner Blue	234G
Windsor Blue	234H
Seacrest Green	234J
Sun Beige	234K
Cream Medium	234L
Live Oak Green Metallic	235A
Grecian Gray	235B
Vista Gray Metallic	235C
Monaco Blue Metallic	235D
Oxford Maroon Metallic	235F
Live Oak Green Metallic/ Grecian Gray	235G
Vista Gray Metallic/ Grecian Gray	235H
Waldorf White	246
Tan Metallic	260

*all except Carryall Suburban
**Sedan Delivery
***Carryall Suburban only

Regular Production Options

1508 Sedan Delivery	$1,465

3100 Series

3102 Chassis w/ flat-face cowl	961
3103 Chassis and cab	1,185
3104 Pickup	1,253
3105 Panel	1,450
3107 Canopy express	1,502
3112 Chassis w/cowl and windshield	982
3116 Carryall suburban w/end-gate	1,700

3600 Series

3602 Chassis w/flat-face cowl	1,060
3603 Chassis and cab	1,284
3604 Pickup	1,372
3608 Chassis and cab w/platform body	1,378
3809 Chassis and cab w/stake body	1,435
3612 Chassis w/cowl and windshield	1,081

3800 Series

3802 Chassis w/flat-face cowl	1,134
3803 Chassis and cab	1,357
3804 Pickup	1,471
3805 Panel	1,669
3807 Canopy express	1,746
3808 Chassis and cab w/platform body	1,487
3809 Chassis and cab w/stake body	1,560

Option Number

200	Double-acting shock absorbers
201	Economy rear axle
204A	Rear axle ratio 5.43:1
208A	Rear axle ratio 5.14:1
210	Rear-view mirrors and brackets, LH/RH
216	Oil-bath air cleaner
227	Heavy-duty clutch
237	Oil filter
241	Governor
249	Tail and stop lamp equipment
256	Heavy duty radiator
263	Right-hand front seat
316	HD 3-speed transmission
318	HD 4-speed transmission
348	Tru-stop brake equipment
386	Chrome radiator grille
387	Rear corner windows
389	Wide running boards
390	Deluxe cab & equipment

1949

Facts

The front inner grille bars were painted silver gray for 1949, making the standard body-colored outer bars stand out more. An additional emblem was located beneath the Chevrolet emblem on each side of the hood. These were This said 3100, 3600, and or 3800 and stood for each model designation.

The foot-operated parking brake was made smaller. The speedometer pointer was painted red, which made it more visible than the previous white one.

The gas tank was relocated behind the driver's seat. This also resulted in the relocation of the fuel filler cap, now at the right side of the cab, behind the door handle.

The cabs used an improved mounting method were attached to the chassis using an improved method incorporating shackle-type mounts.

The 216.5ci engine received some minor changes. The timing gears now received pressurized oil from the front camshaft bearing. Th engine also used 14mm spark plugs, replacing the 10mm plugs.

Of the total calendar year production, 201,537 units were 1/2-tons and 97,668 were 3/4-tons.

The sedan delivery came with the new 1949 passenger-car styling along with passenger-car trim and mechanicals.

Chapter 5

1950 Trucks

Production

Calendar year	Total
1950	494,573

Serial numbers

Description

3HJA00001

3—Assembly Plant Code: 1—Flint MI, 2—Tarrytown NY, 3—St. Louis MO, 5—Kansas City MO, 6—Oakland CA, 8—Atlanta GA, 9—Norwood OH, 14—Baltimore MD, 20—Van Nuys CA, 21—Janesville WI

H—Year, H—1950

J—Series: J—1500, P—3100, R—3600, S—3800

A—Month of production, A—January, B—February, C—March, D—April, E—May, F—June, G—July, H—August, I—September, J—October, K—November, L—December

00001—Consecutive Sequence Number

Model, Wheelbase & GVW

Model Number and Description	Wheelbase (in)	GVW
1500 Series		
1508 Sedan Delivery	115	4,100
3100 Series		
3102 Chassis w/flat-face cowl	116	4,600
3103 Chassis and cab	116	4,600
3104 Pickup	116	4,600
3105 Panel	116	4,600
3107 Canopy express	116	4,600
3112 Chassis w/cowl and windshield	116	4,600
3116 Carryall suburban w/end-gate	116	4,600
3122 Stripped chassis w/flat-face cowl	116	4,600
3132 Stripped chassis w/cowl and windshield	116	4,600
3600 Series		
3602 Chassis w/flat-face cowl	125¼	5,800
3603 Chassis and cab	125¼	5,800
3604 Pickup	125¼	5,800
3608 Chassis and cab w/platform body	125¼	5,800
3809 Chassis and cab w/stake body	125¼	5,800

3612	Chassis w/cowl and windshield	125¼	5,800
3622	Stripped chassis w/flat-face cowl	125¼	5,800
3632	Stripped chassis w/cowl and windshield	125¼	5,800

3800 Series

3802	Chassis w/flat-face cowl	137	8,800
3803	Chassis and cab	137	8,800
3804	Pickup	137	6,700
3805	Panel	137	6,700
3807	Canopy express	137	6,700
3808	Chassis and cab w/platform body	137	8,800
3809	Chassis and cab w/stake body	137	8,800
3812	Chassis w/cowl and windshield	137	8,800
3822	Stripped chassis w/flat-face cowl	137	8,800
3832	Stripped chassis w/cowl and windshield	137	8,800

Engine & Transmission Suffix Codes

1500 Series

HAA—216.5ci I-6 1bbl 92hp—three-speed manual
HAM—216.5ci I-6 1bbl 92hp—three-speed manual
HAC—216.5ci I-6 1bbl 92hp—three-speed manual
HAP—216.5ci I-6 1bbl 92hp—three-speed manual

3100 Series

HBA—216.5ci I-6 1bbl 92hp—three/four speed manual
HBM—216.5ci I-6 1bbl 92hp—three/four speed manual
BHCA—216.5ci I-6 1bbl 92hp—three/four speed manual*
BHCM—216.5ci I-6 1bbl 92hp—three/four speed manual*

3600/3800 Series

AHCF—216.5ci I-6 1bbl 92hp—three/four speed manual*
AHCM—216.5ci I-6 1bbl 92hp—three/four speed manual*
*"A" will precede prefix if engine is used on 3600/3800 Series
*"B" will precede prefix if engine is used on 3100 Series

Transmission Codes

Code	Type	Plant
1500 Series		
HA	3spd	Saginaw
HB	3spd	Muncie
HC	3spd	Toledo
HG	3spd	Saginaw
HH	3spd	Muncie
HJ	3spd	Toledo
3100 Series		
HN	3spd	Saginaw
HO	3spd	Muncie
HP	3spd	Toledo
3600 Series		
HQ	3spd	Saginaw
HR	3spd	Muncie
HS	3spd	Toledo
3800 Series		
TA	4spd	Saginaw
TB	4spd	Muncie
TC	4spd	Toledo
3100/3600 Series		
TK	4spd	Saginaw
TL	4spd	Muncie
TM	4spd	Saginaw

Axle Identification

Code	Ratio	Plant
1500 Series		
HA	4.11:1	G&A
HB	4.11:1	Buffalo

3100 Series

HE	4.11:1	G&A
HF	4.11:1	Buffalo

3600 Series

HG	4.57:1	G&A
HH	4.57:1	Buffalo

3800 Series

TJ	5.14:1	G&A
TK	5.14:1	Buffalo
TQ	5.14:1 (dual wheels)	G&A
TR	5.14:1 (dual wheels)	Buffalo

Exterior Color Codes

Forester Green	STD
Maryland Black	STD*
Swift Red	234A
Armour Yellow	234B
White	234C
Jet Black	234D
Omaha Orange	234E
Cape Maroon	234F
Mariner Blue	234G
Windsor Blue	234H
Seacrest Green	234J
Sun Beige	234K
Cream Medium	234L
Mist Green	235A
Grecian Gray	235B
Falcon Green	235C
Windsor Blue	235D
Oxford Maroon	235F
Mist Green/Crystal Green	235G
Grecian Gray/Falcon Gray	235H

*Sedan Delivery

Regular Production Options

1508 Sedan Delivery	$1,455

3100 Series

3102 Chassis w/ flat-face cowl	951
3103 Chassis and cab	1,175
3104 Pickup	1,243
3105 Panel	1,440
3106 Carryall suburban w/panel rear doors	1,690
3107 Canopy express	1,492
3112 Chassis w/cowl and windshield	972
3116 Carryall suburban w/end-gate	1,690

3600 Series

3602 Chassis w/flat-face cowl	1,050
3603 Chassis and cab	1,282
3604 Pickup	1,362
3608 Chassis and cab w/platform body	1,368
3809 Chassis and cab w/stake body	1,425
3612 Chassis w/cowl and windshield	1,071

3800 Series

3802 Chassis w/flat-face cowl	1,124
3803 Chassis and cab	1,347
3804 Pickup	1,461
3805 Panel	1,659
3807 Canopy express	1,736
3808 Chassis and cab w/platform body	1,477
3809 Chassis and cab w/stake body	1,550

Option Number

200 Double-acting shock absorbers
207 Long running boards and rear fenders
208 Rear axle ratio 5.14:1
210 Rear-view mirrors and brackets, LH/RH
211 Rear shock absorber shields
213 Hydrovac power brake
216 Oil-bath air cleaner
227 Heavy-duty clutch
237 Oil filter
241 Governor
249 Dual tail and stop lights
254 Heavy duty rear springs
256 Heavy duty radiator
263 Right-hand front seat
267 Auxiliary rear springs
316 HD 3-speed transmission
318 HD 4-speed transmission
340 Combination fuel and vacuum pump
361 Genuine leather trim
367 Front bumper (Forward-Control chassis)
386 Chrome radiator grille
387 Rear corner windows
390 Deluxe equipment cab & panel

Facts

The Suburban could be had with rear panel-type doors in addition to or the liftgate. It now also used the same colors as other Chevrolet trucks.

All 3000 Series trucks used tube-type shock absorbers, replacing a lever-action type.

The engine received minor improvements, resulting in a 92hp rating. Replacing the previous Carter carburetor was a GM model B unit.

The voltage regulator was relocated to the left side of the cowl; it had been mounted on the left fender skirt.

In the interior, the seat was slightly wider than before.

Chapter 6

1951 Trucks

Production

Sedan Delivery	20,817
Calendar year	Total
1951	426,115

Serial numbers

Description

3JJA00001

3—Assembly Plant Code: 1—Flint MI, 2—Tarrytown NY, 3—St. Louis MO, 5—Kansas City MO, 6—Oakland CA, 8—Atlanta GA, 9—Norwood OH, 14—Baltimore MD, 20—Van Nuys CA, 21—Janesville WI

J—Year, J—1951

J—Series: J—1500, P—3100, R—3600, S—3800

A—Month of production, A—January, B—February, C—March, D—April, E—May, F—June, G—July, H—August, I—September, J—October, K—November, L—December

00001—Consecutive Sequence Number

Model, Wheelbase & GVW

Model Number and Description	Wheelbase (in)	GVW
1500 Series		
1508 Sedan Delivery	115	4,100
3100 Series		
3102 Chassis w/flat-face cowl	116	4,600
3103 Chassis and cab	116	4,600
3104 Pickup	116	4,600
3105 Panel	116	4,600
3106 Carryall suburban w/panel type rear doors	116	4,600
3107 Canopy express	116	4,600
3112 Chassis w/cowl and windshield	116	4,600
3116 Carryall suburban w/end-gate	116	4,600
3600 Series		
3602 Chassis w/flat-face cowl	125¼	5,800
3603 Chassis and cab	125¼	5,800
3604 Pickup	125¼	5,800
3608 Chassis and cab w/platform body	125¼	5,800
3809 Chassis and cab w/stake body	125¼	5,800
3612 Chassis w/cowl and windshield	125¼	5,800

3800 Series

3802	Chassis w/flat-face cowl	137	7,000
3803	Chassis and cab	137	7,000
3804	Pickup	137	7,000
3805	Panel	137	7,000
3807	Canopy express	137	7,000
3808	Chassis and cab w/platform body	137	7,000
3809	Chassis and cab w/stake body	137	7,000
3812	Chassis w/cowl and windshield	137	7,000

Engine & Transmission Suffix Codes

1500 Series

JAA—216.5ci I-6 1bbl 92hp—three-speed manual
JAM—216.5ci I-6 1bbl 92hp—three-speed manual
JAC—216.5ci I-6 1bbl 92hp—three-speed manual
JAP—216.5ci I-6 1bbl 92hp—three-speed manual

3100 Series

JBA—216.5ci I-6 1bbl 92hp—three/four speed manual
JBM—216.5ci I-6 1bbl 92hp—three/four speed manual
BJCA—216.5ci I-6 1bbl 92hp—three/four speed manual
BJCM—216.5ci I-6 1bbl 92hp—three/four speed manual

3600 Series

AJCA—216.5ci I-6 1bbl 92hp—three/four speed manual
N/AAJCM—216.5ci I-6 1bbl 92hp—three/four speed manual

3600/3800 Series

JCD*—216.5ci I-6 1bbl 92hp—three/four speed manual*
JCQ*—216.5ci I-6 1bbl 92hp—three/four speed manual*
"A" will precede prefix if engine is used on 3600/3800 Series
"B" will precede prefix if engine is used on 3100 Series
*with Hydrovac power brakes

Transmission Codes

Code	Type	Plant
1500 Series		
JA	3spd	Saginaw
JB	3spd	Muncie
JC	3spd	Toledo
JG	3spd	Saginaw
JH	3spd	Muncie
JJ	3spd	Toledo
3100 Series		
JN	3spd	Saginaw
JO	3spd	Muncie
JP	3spd	Toledo
3600 Series		
JQ	3spd	Saginaw
JR	3spd	Muncie
JS	3spd	Toledo

3800 Series

UA	4spd	Saginaw
UB	4spd	Muncie
UC	4spd	Toledo

3100/3600 Series

UK	4spd	Saginaw
UL	4spd	Muncie
UM	4spd	Saginaw

Axle Identification

Code	Ratio	Plant
1500 Series		
JA	4.11:1	G&A
JB	4.11:1	Buffalo
3100 Series		
JE	4.11:1	G&A
JF	4.11:1	Buffalo

3600 Series

JG	4.57:1	G&A
JH	4.57:1	Buffalo
JQ	5.14:1	G&A
JR	5.14:1	Buffalo

3800 Series

UJ	5.14:1	G&A
UK	5.14:1	Buffalo
UQ	5.14:1 (dual wheels)	G&A
UR	5.14:1 (dual wheels)	Buffalo

Exterior Color Codes

Forester Green	STD
Maryland Black	STD*
Swift Red	234A
Armour Yellow	234B
White	234C
Jet Black	234D
Omaha Orange	234E
Cape Maroon	234F
Mariner Blue	234G
Windsor Blue	234H
Seacrest Green	234J
Sun Beige	234K
Cream Medium	234L
Aspen Green	235A
Thistle Gray	235B
Trophy Blue	235D
Burgundy Red	235F
Waldorf White	246
*Sedan Delivery	

Regular Production Options

1508 Sedan Delivery	$1,455

3100 Series

3102 Chassis w/ flat-face cowl	1,035
3103 Chassis and cab	1,282
3104 Pickup	1,353
3105 Panel	1,556
3106 Carryall suburban w/panel type rear doors	1,818
3107 Canopy express	1,610
3112 Chassis w/cowl and windshield	1,057
3116 Carryall suburban w/end-gate	1,818

3600 Series

3602 Chassis w/flat-face cowl	1,170
3603 Chassis and cab	1,417
3604 Pickup	1,508
3608 Chassis and cab w/platform body	1,514
3809 Chassis and cab w/stake body	1,578
3612 Chassis w/cowl and windshield	1,190

3800 Series

3802 Chassis w/flat-face cowl	1,124
3803 Chassis and cab	1,347
3804 Pickup	1,461
3805 Panel	1,659
3807 Canopy express	1,736
3808 Chassis and cab w/platform body	1,477
3809 Chassis and cab w/stake body	1,550

Option Number

200 Double-acting shock absorbers
207 Long running boards and rear fenders
208 Rear axle ratio 5.14:1
210 Rear-view mirrors and brackets, LH/RH
211 Rear shock absorber shields
213 Hydrovac power brake
216 Oil-bath air cleaner
218 Rear bumper
227 Heavy-duty clutch
237 Oil filter
241 Governor
249 Dual tail and stop lights
254 Heavy duty rear springs
256 Heavy duty radiator
263 Auxiliary seat
267 Auxiliary rear springs
281 Vacuum Reserve Tank
316 HD 3-speed transmission
318 HD 4-speed transmission
326 High output generator
328 Stand-drive controls
340 Combination fuel and vacuum pump
367 Front bumper
384 Spare wheel & carrier
386 Chrome radiator grille
387 Rear corner windows
390 Deluxe equipment cab & panel

Facts

The most noticeable exterior change in 1951 was the use of side vent windows. To accommodate these, the door glass was reduced in size. In the interior, different seat adjusters were used. The rear bumper, which had been standard equipment on 1/2- and 3/4-ton models, became optional.

The brakes were improved on the 3100 Series trucks. The brakes were self-energizing, which resulted in improved brake action. The front stabilizer bar and the procedures for mounting the shock absorbers were also improved.

Chapter 7

1952 Trucks

Production

Sedan Delivery	9,715
Calendar year	Total
1952	272,249

Serial numbers

Description

3KJA00001

3—Assembly Plant Code: 1—Flint MI, 2—Tarrytown NY, 3—St. Louis MO, 5—Kansas City MO, 6—Oakland CA, 8—Atlanta GA, 9—Norwood OH, 14—Baltimore MD, 20—Van Nuys CA, 21—Janesville WI

K—Year, K—1952

J—Series: J—1500, P—3100, R—3600, S—3800

A—Month of production, A—January, B—February, C—March, D—April, E—May, F—June, G—July, H—August, I—September, J—October, K—November, L—December

00001—Consecutive Sequence Number

Model, Wheelbase & GVW

Model Number and Description	Wheelbase (in)	GVW
1500 Series		
1508 Sedan Delivery	115	4,100
3100 Series		
3102 Chassis w/flat-face cowl	116	4,800
3103 Chassis and cab	116	4,800
3104 Pickup	116	4,800
3105 Panel	116	4,800
3106 Carryall suburban w/panel type rear doors	116	4,800
3107 Canopy express	116	4,800
3112 Chassis w/cowl and windshield	116	4,800
3116 Carryall suburban w/end-gate	116	4,800
3600 Series		
3602 Chassis w/flat-face cowl	125¼	5,800
3603 Chassis and cab	125¼	5,800
3604 Pickup	125¼	5,800
3608 Chassis and cab w/platform body	125¼	5,800
3809 Chassis and cab w/stake body	125¼	5,800
3612 Chassis w/cowl and windshield	125¼	5,800

3800 Series

3802 Chassis w/flat-face cowl	137	7,000
3803 Chassis and cab	137	7,000
3804 Pickup	137	7,000
3805 Panel	137	7,000
3807 Canopy express	137	7,000
3808 Chassis and cab w/platform body	137	7,000
3809 Chassis and cab w/stake body	137	7,000
3812 Chassis w/cowl and windshield	137	7,000

Engine & Transmission Suffix Codes

1500 Series

KAA—216.5ci I-6 1bbl 92hp—three-speed manual
KAM—216.5ci I-6 1bbl 92hp—three-speed manual
KAC—216.5ci I-6 1bbl 92hp—three-speed manual
KAP—216.5ci I-6 1bbl 92hp—three-speed manual

3100 Series

KBA—216.5ci I-6 1bbl 92hp—three/four speed manual
KBM—216.5ci I-6 1bbl 92hp—three/four speed manual
BKCA—216.5ci I-6 1bbl 92hp—three/four speed manual
BKCM—216.5ci I-6 1bbl 92hp—three/four speed manual

3600/3800 Series

AKCA—216.5ci I-6 1bbl 92hp—three/four speed manual
AKCM—216.5ci I-6 1bbl 92hp—three/four speed manual
AKCD—216.5ci I-6 1bbl 92hp—three/four speed manual*
"A" will precede prefix if engine is used on 3600/3800 Series
"B" will precede prefix if engine is used on 3100 Series
*with Hydrovac power brakes

Transmission Codes

Code	Type	Plant
1500 Series		
KA	3spd	Saginaw
KB	3spd	Muncie
KC	3spd	Toledo
KG	3spd	Saginaw
KH	3spd	Muncie
KJ	3spd	Toledo
3100 Series		
KN	3spd	Saginaw
KO	3spd	Muncie
KP	3spd	Toledo
3600 Series		
KQ	3spd	Saginaw
KR	3spd	Muncie
KS	3spd	Toledo

3800 Series

VA	4spd	Saginaw
VB	4spd	Muncie
VC	4spd	Toledo

3100/3600 Series

VK	4spd	Saginaw
VL	4spd	Muncie
VM	4spd	Saginaw

Axle Identification

Code	Ratio	Plant
1500 Series		
KA	4.11:1	G&A
KB	4.11:1	Buffalo
3100 Series		
KE	4.11:1	G&A
KF	4.11:1	Buffalo

3600 Series

KG	4.57:1	G&A
KH	4.57:1	Buffalo
KQ	5.14:1	G&A
KR	5.14:1	Buffalo

3800 Series

VJ	5.14:1	G&A
VK	5.14:1	Buffalo
VQ	5.14:1	G&A
VR	5.14:1	Buffalo

Exterior Color Codes

Forester Green	STD
Onyx Black	STD*
Swift Red	234A
Armour Yellow	234B
White	234C
Jet Black	234D
Omaha Orange	234E
Cape Maroon	234F
Mariner Blue	234G
Windsor Blue	234H
Seacrest Green	234J
Sun Beige	234K
Cream Medium	234L
Birch Gray	235A
Dusk Gray	235B
Spring Green	235C
Emerald Green	235D
Admiral Blue	235F
Sahara Beige	235G
Regal Maroon	235H
Waldorf White	246
*Sedan Delivery	

Regular Production Options

| 1508 Sedan Delivery | $1,648 |

3100 Series

3102 Chassis w/ flat-face cowl	1,076
3103 Chassis and cab	1,334
3104 Pickup	1,407
3105 Panel	1,619
3106 Carryall suburban w/panel type rear doors	1,933
3107 Canopy express	1,675
3112 Chassis w/cowl and windshield	1,098
3116 Carryall suburban w/end-gate	1,933

3600 Series

3602 Chassis w/flat-face cowl	1,216
3603 Chassis and cab	1,474
3604 Pickup	1,658
3608 Chassis and cab w/platform body	1,575
3809 Chassis and cab w/stake body	1,642
3612 Chassis w/cowl and windshield	1,237

3800 Series

3802 Chassis w/flat-face cowl	1,312
3803 Chassis and cab	1,570
3804 Pickup	1,692
3805 Panel	1,916
3807 Canopy express	2,000
3808 Chassis and cab w/platform body	1,709
3809 Chassis and cab w/stake body	1,782
3812 Chassis w/cowl and windshield	1,335

Option Number

200	Double-acting shock absorbers
207	Long running boards and rear fenders
208	Rear axle ratio 5.14:1
210	Rear-view mirrors and brackets, LH/RH
211	Rear shock absorber shields
213	Hydrovac power brake
216	Oil-bath air cleaner
217	Positive crankcase ventilation
218	Rear bumper
227	Heavy-duty clutch
237	Oil filter
241	Governor
249	Dual tail and stop lights
254	Heavy duty rear springs
256	Heavy duty radiator
263	Auxiliary seat
267	Auxiliary rear springs
281	Vacuum Reserve Tank
316	HD 3-speed transmission
318	HD 4-speed transmission
326	High output generator
327	Solenoid starter
328	Stand-drive controls
340	Combination fuel and vacuum pump
384	Spare wheel & carrier
387	Rear corner windows
395	Left hand key lock

1952. Chevrolet built some handsome . . . .this era too. Applegate & Applegate

Facts

The exterior door handles were changed from a turn-handle type to a push-button type in 1952. A lockable door on the left side was optional.

A running production change was the deletion of the Series emblems as well as the Deluxe emblem on trucks so equipped. The 3/4-ton models still came with 3600 emblems.

All models used a pressurized cooling system that incorporated 3-1/2lb to 4-1/2lb radiator caps.

In the interior, the speedometer indicated 90mph as the top speed and the high beam indicator was relocated to the top. A round plastic knob was used instead of the previous T-handle on the brake release lever.

Chapter 8

1953 Trucks

Production

Sedan Delivery	15,523
Calendar year	Total
1953	361,833

Serial numbers

Description

A3D000001

A—Assembly Plant Code: A—Atlanta GA, B—Baltimore, F—Flint MI,
   J—Janesville WI, K—Kansas City MO, L—Los Angeles CA, N—Norwood
   OH, O—Oakland CA, S—St. Louis MO, T—Tarrytown NY

3—Year, 3—1953

D—Series: D—1500, H—3100, J—3600, L—3800

000001—Consecutive Sequence Number

Model, Wheelbase & GVW

Model Number and Description	Wheelbase (in)	GVW
1500 Series		
1508 Sedan Delivery	115	4,100
3100 Series		
3102 Chassis w/flat-face cowl	116	4,800
3103 Chassis and cab	116	4,800
3104 Pickup	116	4,800
3105 Panel	116	4,800
3106 Carryall suburban w/panel type rear doors	116	4,800
3107 Canopy express	116	4,800
3112 Chassis w/cowl and windshield	116	4,800
3116 Carryall suburban w/end-gate	116	4,800
3600 Series		
3602 Chassis w/flat-face cowl	125¼	5,800
3603 Chassis and cab	125¼	5,800
3604 Pickup	125¼	5,800
3608 Chassis and cab w/platform body	125¼	5,800
3809 Chassis and cab w/stake body	125¼	5,800
3612 Chassis w/cowl and windshield	125¼	5,800

3800 Series

3802	Chassis w/flat-face cowl	137	8,800
3803	Chassis and cab	137	8,800
3804	Pickup	137	7,000
3805	Panel	137	7,000
3807	Canopy express	137	7,000
3808	Platform (stake pocket rear)	137	8,800
3809	Chassis and cab w/stake body	137	8,800
3812	Chassis w/cowl and windshield	137	8,800

Engine & Transmission Suffix Codes

1500 Series
LAG—216.5ci I-6 1bbl 92hp—three-speed manual
LAT—216.5ci I-6 1bbl 92hp—three-speed manual
LAJ—216.5ci I-6 1bbl 92hp—three-speed manual
LAV—216.5ci I-6 1bbl 92hp—three-speed manual

3100 Series
LBA—216.5ci I-6 1bbl 92hp—three/four speed manual
LBM—216.5ci I-6 1bbl 92hp—three/four speed manual
BLCC—216.5ci I-6 1bbl 92hp—three/four speed manual
BLCP—216.5ci I-6 1bbl 92hp—three/four speed manual

3600 Series
LCC—216.5ci I-6 1bbl 92hp—three/four speed manual

3800 Series
ALCC—216.5ci I-6 1bbl 92hp—three/four speed manual
ALCP—216.5ci I-6 1bbl 92hp—three/four speed manual
LCH—216.5ci I-6 1bbl 92hp—three/four speed manual*
LCU—216.5ci I-6 1bbl 92hp—three/four speed manual*
*with positive crankcase ventilation

Transmission Codes

Code	Type	Plant
1500 Series		
LA	3spd	Saginaw
LB	3spd	Muncie
LC	3spd	Toledo
LG	3spd	Saginaw
LH	3spd	Muncie
LJ	3spd	Toledo
3100 Series		
LN	3spd	Saginaw
LO	3spd	Muncie
LP	3spd	Toledo
3600 Series		
LQ	3spd	Saginaw
LR	3spd	Muncie
LS	3spd	Toledo

3800 Series

WD	4spd	Saginaw
WE	4spd	Muncie
WF	4spd	Toledo

3100/3600 Series

WK	4spd	Saginaw
WL	4spd	Muncie
WM	4spd	Saginaw

Axle Identification

Code	Ratio	Plant
1500 Series		
LA	4.11:1	G&A
LB	4.11:1	Buffalo
3100 Series		
LU	4.11:1	G&A
LV	4.11:1	Buffalo

3600 Series

LG	4.57:1	G&A
LH	4.57:1	Buffalo
LQ	5.14:1	G&A
LR	5.14:1	Buffalo

3800 Series

WJ	5.14:1	G&A
WK	5.14:1	Buffalo
WQ	5.14:1	G&A
WR	5.14:1	Buffalo

Exterior Color Codes

Juniper Green	STD
Onyx Black	STD*
Thistle Gray	094
Monotone Hysheen Gray	202
Driftwood Gray	231A
Dusk Gray	231B
Woodland Green	231C
Surf Green	231D
Regatta Blue	231E
Sahara Beige	231F
Madeira Maroon	231G
Commercial Red	234A
Jet Black	234D
Cream Medium	234L
Yukon Yellow	234M
Ocean Green	234N
Transport Blue	234P
Burgundy Maroon	234Q
Coppertone	234R
Autumn Brown	234S
Pure White	234T
*Sedan Delivery	

Regular Production Options

1508 Sedan Delivery	$1,648

3100 Series

3102 Chassis w/ flat-face cowl	1,076
3103 Chassis and cab	1,334
3104 Pickup	1,407
3105 Panel	1,620
3106 Carryall suburban w/panel type rear doors	1,947
3107 Canopy express	1,676
3112 Chassis w/cowl and windshield	1,099
3116 Carryall suburban w/end-gate	1,947

3600 Series

3602 Chassis w/flat-face cowl	1,216
3603 Chassis and cab	1,474
3604 Pickup	1,569
3608 Chassis and cab w/platform body	1,575
3809 Chassis and cab w/stake body	1,642
3612 Chassis w/cowl and windshield	1,238

3800 Series

3802 Chassis w/flat-face cowl	1,312
3803 Chassis and cab	1,570
3804 Pickup	1,692
3805 Panel	1,916
3807 Canopy express	2,000
3808 Platform (stake pocket rear)	1,709
3809 Chassis and cab w/stake body	1,782
3812 Chassis w/cowl and windshield	1,335

Option Number

200	Double-acting shock absorbers
207	Long running boards and rear fenders
208	Rear axle ratio 5.14:1
210	Rear-view mirrors and brackets, LH/RH
211	Rear shock absorber shields
213	Hydrovac power brake
214	Propeller shaft brake
216	Oil-bath air cleaner
217	Positive crankcase ventilation
218	Rear bumper
227	Heavy-duty clutch
237	Oil filter
241	Governor
249	Dual tail and stop lights
254	Heavy duty rear springs
256	Heavy duty radiator
263	Auxiliary seat
267	Auxiliary rear springs
281	Vacuum Reserve Tank
316	HD 3-speed transmission
318	HD 4-speed transmission
325,326	High output generator
327	Solenoid starter
340	Combination fuel and vacuum pump
341	Side mounted wheel carrier
387	Rear corner windows
395	Left hand key lock
396	Spare wheel lock equipment
399	E-Z Eye glass

This 1953 Chevrolet promotional photograph features a pickup with the big rear window, and lots of chrome trim. Applegate & Applegate

Facts

New colors were made available for the 1953 models, with Juniper Green becoming the standard color. Complementing the new green was Cream Medium striping. New options included tinted windows, a side-mounted spare tire carrier, and a rear bumper on the 1/2- and 3/4-ton models. This was also the first year for factory-installed turn signals, which were optional according to the factory but mandatory in certain states.

New, too, were the hoodside emblems, which placed the series designation numbers above a chrome molding.

Not as visible were numerous mechanical improvements, which included a new type of wheel bearing on the 3/4-ton models. Altered wheel alignment settings were seen on the 1/2- and 3/4-ton models.

In the interior, a new instrument panel incorporated a 60lb psu oil pressure gauge. Also new were interior colors, interior fabrics, and a larger steering wheel. A window crank mechanism opened and closed the side vent windows.

The sedan delivery used the new 1953 Chevrolet passenger-car styling. New styling features included a one-piece windshield. The sedan delivery was part of the base 150 Series and did not have any bodyside moldings. The standard 235ci six was rated 108hp with the manual three-speed and at 115hp with the Powerglide two-speed automatic. A new option was power steering.

Chapter 9

1954 Trucks

Production

Sedan Delivery	8,255
Calendar year	Total
1954	325,515

Serial numbers

Description

D54A000001

D—Series: D—1500, H—3100, J—3600, L—3800

54—Year, 54—1954

A—Assembly Plant Code: A—Atlanta GA, B—Baltimore MD, F—Flint MI, J—Janesville WI, K—Kansas City MO, L—Los Angeles CA, N—Norwood OH, O—Oakland CA, S—St. Louis MO, T—Tarrytown NY

000001—Consecutive Sequence Number

Model, Wheelbase & GVW

Model Number and Description	Wheelbase (in)	GVW
1500 Series		
1508 Sedan delivery	115	4,100
3100 Series		
3102 Chassis w/flat-face cowl	116	4,800
3103 Chassis and cab	116	4,800
3104 Pickup	116	4,800
3105 Panel	116	4,800
3106 Carryall suburban w/panel type rear doors	116	4,800
3107 Canopy express	116	4,800
3112 Chassis w/cowl and windshield	116	4,800
3116 Carryall suburban w/end-gate	116	4,800
3600 Series		
3602 Chassis w/flat-face cowl	125¼	5,800
3603 Chassis and cab	125¼	5,800
3604 Pickup	125¼	5,800
3608 Chassis and cab w/platform body	125¼	5,800
3809 Chassis and cab w/stake body	125¼	5,800
3612 Chassis w/cowl and windshield	125¼	5,800

3800 Series

3802	Chassis w/flat-face cowl	137	8,800
3803	Chassis and cab	137	8,800
3804	Pickup	137	7,000
3805	Panel	137	7,000
3807	Canopy express	137	7,000
3808	Platform (stake pocket rear)	137	8,800
3809	Chassis and cab w/stake body	137	8,800
3812	Chassis w/cowl and windshield	137	8,800

Engine & Transmission Suffix Codes

1500 Series

F54Z,T54Z—235ci I-6 1bbl 123hp—manual trans
F54ZC,T54ZC—235ci I-6 1bbl 123hp—manual trans HD
F54ZE,T54ZE—235ci I-6 1bbl 123hp—manual trans
F54Y,T54Y—235ci I-6 1bbl 123hp—Powerglide automatic
F54YE,T54YE—235ci I-6 1bbl 123hp—Powerglide automatic

3100 Series

F54X,T54X—235ci I-6 1bbl 112hp—manual trans
F54U,T54U—235ci I-6 1bbl 112hp—manual trans HD
F54M,T54M—235ci I-6 1bbl 112hp—automatic trans

3600 Series

F54XA,T54XA—235ci I-6 1bbl 112hp—manual trans
F54UA,T54UA—235ci I-6 1bbl 112hp—manual trans HD
F54MA,T54MA—235ci I-6 1bbl 112hp—automatic trans

3800 Series

F54XB,T54XB—235ci I-6 1bbl 112hp—manual trans
F54UB,T54UB—235ci I-6 1bbl 112hp—manual trans HD
F54MB,T54MB—235ci I-6 1bbl 112hp—automatic trans

Transmission Codes

Code	Type	Plant
M	3 speed	Muncie
S	3 speed	Saginaw
W	3 speed HD	Warner Gear
T	4 speed	Toledo

Axle Identification

Code	Ratio	Plant

1500 Series

ML	3.70:1	G&A
MM	3.70:1	Buffalo
MS	3.55:1	G&A
MT	3.55:1	Buffalo
ME	4.11:1	G&A
MF	4.11:1	Buffalo

3100 Series

MU	3.90:1	G&A
MV	3.90:1	Buffalo
MA	3.90:1	G&A
MB	3.90:1	Buffalo

3600 Series

MG	4.57:1	G&A
MH	4.57:1	Buffalo
MD	4.57:1	G&A
MC	5.14:1	G&A
MQ	5.14:1	G&A
MR	5.14:1	Buffalo

3800 Series

XJ	5.14:1	G&A
XK	5.14:1	Buffalo
XQ	5.14:1	G&A
XR	5.14:1	Buffalo

Exterior Color Codes

Juniper Green	STD
Commercial Red	234A
Jet Black	234D
Mariner Blue	234G
Cream Medium	234L
Yukon Yellow	234M
Ocean Green	234N
Transport Blue	234P
Burgundy Maroon	234Q
Coppertone	234R
Autumn Brown	234S
Pure White	234T

Regular Production Options

1508 Sedan Delivery	$1,632

3100 Series

3102 Chassis w/ flat-face cowl	1,087
3103 Chassis and cab	1,346
3104 Pickup	1,419
3105 Panel	1,631
3106 Carryall suburban w/panel type rear doors	1,958
3107 Canopy express	1,688
3112 Chassis w/cowl and windshield	1,109
3116 Carryall suburban w/end-gate	1,958

3600 Series

3602 Chassis w/flat-face cowl	1,227
3603 Chassis and cab	1,486
3604 Pickup	1,582
3608 Chassis and cab w/platform body	1,587
3809 Chassis and cab w/stake body	1,654
3612 Chassis w/cowl and windshield	1,249

3800 Series

3802 Chassis w/flat-face cowl	1,325
3803 Chassis and cab	1,582
3804 Pickup	1,705
3805 Panel	1,929
3807 Canopy express	2,012
3808 Platform (stake pocket rear)	1,722
3809 Chassis and cab w/stake body	1,794
3812 Chassis w/cowl and windshield	1,347

Option Number

100	Directional signals
200	Double-acting shock absorbers
207	Long running boards and rear fenders
208	Rear axle ratio 5.14:1
210	Rear-view mirrors and brackets, LH/RH
211	Rear shock absorber shields
213	Hydrovac power brake
214	Propeller shaft brake
216	Oil-bath air cleaner
217	Positive crankcase ventilation
218	Rear bumper
227	Heavy-duty clutch
230	High sill body
237	Oil filter
241	Governor
249	Dual tail and stop lights
254	Heavy duty rear springs
256	Heavy duty radiator
263	Auxiliary seat
267	Auxiliary rear springs
281	Vacuum Reserve Tank
314	Hydramatic 4-speed automatic transmission
315	Overdrive transmission
316	HD 3-speed transmission
318	HD 4-speed transmission
320	Constant speed electric windshield wiper
324	Power steering
325,326	High output generator
327	Solenoid starter
340	Combination fuel and vacuum pump
341	Side mounted wheel carrier
367	Front bumper
384	Spare wheel & carrier
387	Rear corner windows
390	Deluxe equipment
393	Chrome equipment
395	Left hand door and side mounted spare wheel key lock
396	Spare wheel lock equipment
399	E-Z Eye glass
412	Power brakes
417	Positive crankcase ventilation
430,431	Deluxe cabs & panels
438	Two-tone cab exterior paint color
439	Two-tone panel exterior paint color

1954 3100 1/2 ton.

Facts

The 1954 models got a new body-colored grille with three horizontal bars, body colored grille and a single-piece windshield. C-h-e-v-r-o-l-e-t block letters were stamped on the center of the top bar. Above the grille was a new bow-tie hood emblem. Optional was a chrome-plated grille, but this grille was not part of the Deluxe Package. Deluxe-equipped trucks came with painted bumpers, a painted grille, and painted hubcaps. The hubcaps were new and featured a bow-tie design in the center.

A larger version of the Chevrolet six-cylinder engine, displacing 235ci, was used. Its output was 112hp. A noteworthy feature was its full-pressure oiling system. A heavier rear engine cross-member increased frame rigidity.

The pickup box was also redesigned and featured a 2in-lower loading height, flat-top side panels, and a new tailgate latch. This box design continued in the same basic form until 1988.

Juniper Green was still the standard color, and others were optional. Extracost were two-tone color combinations. On these, the upper body was painted Shell White. Wheels were painted black on all models except the Deluxe, where they were body colored. Two-tone trucks came with wheels painted in the lower-body color painted wheels with triple stripes.

The standard cloth seat covering in the cab came in either brown or gray/maroon. Deluxe cloth interiors were finished in maroon/beige, brown/white, or green. Also included in the Deluxe cab option were stainless steel windshield and side window reveals, chrome vent window frames, a right-hand visor, a left-side armrest, a cigar lighter, dual horns, and rear corner windows.

New on the option list was the Hydra-matic four speed automatic. Also new was a heavy-duty three-speed manual from Warner Gear.

Chapter 10

1955 Trucks

Production

Calendar year	Total
1955	393,312*

Includes 1955 Second Series

Serial numbers

Description

D55A000001

D—Series: D—1500, H—3100, J—3600, L—3800

55—Year, 55—1955

A—Assembly Plant Code: A—Atlanta GA, B—Baltimore MD, F—Flint MI,
J—Janesville WI, K—Kansas City MO, L—Los Angeles CA, N—Norwood
OH, O—Oakland CA, S—St. Louis MO, T—Tarrytown NY

000001—Consecutive Sequence Number

Model, Wheelbase & GVW

Model Number and Description	Wheelbase (in)	GVW
1500 Series		
1508 Sedan Delivery	115	4,100
3100 Series		
3102 Chassis w/flat-face cowl	116	4,800
3103 Chassis and cab	116	4,800
3104 Pickup	116	4,800
3105 Panel	116	4,800
3106 Carryall suburban w/panel type rear doors	116	4,800
3107 Canopy express	116	4,800
3112 Chassis w/cowl and windshield	116	4,800
3116 Carryall suburban w/end-gate	116	4,800
3600 Series		
3602 Chassis w/flat-face cowl	125¼	5,800
3603 Chassis and cab	125¼	5,800
3604 Pickup	125¼	5,800
3608 Chassis and cab w/platform body	125¼	5,800
3809 Chassis and cab w/stake body	125¼	5,800
3612 Chassis w/cowl and windshield	125¼	5,800

3800 Series

3802	Chassis w/flat-face cowl	137	8,800
3803	Chassis and cab	137	8,800
3804	Pickup	137	7,000
3805	Panel	137	7,000
3807	Canopy express	137	7,000
3808	Platform		
3809	Chassis and cab w/stake body	137	8,800
3812	Chassis w/cowl and windshield	137	8,800

Engine & Transmission Suffix Codes

1500 Series
FZ—235ci I-6 1bbl 123hp—manual trans
TZ—235ci I-6 1bbl 123hp—manual trans
FZC—235ci I-6 1bbl 123hp—manual trans
TZC—235ci I-6 1bbl 123hp—manual trans
FZE—235ci I-6 1bbl 123hp—manual trans
TZE—235ci I-6 1bbl 123hp—manual trans
FY—235ci I-6 1bbl 123hp—auto trans
TY—235ci I-6 1bbl 123hp—auto trans
FYE—235ci I-6 1bbl 123hp—auto trans
TYE—235ci I-6 1bbl 123hp—auto trans

3100 Series
FX—235ci I-6 1bbl 112hp—manual trans
TX—235ci I-6 1bbl 112hp—manual trans
FU—235ci I-6 1bbl 112hp—manual trans
TU—235ci I-6 1bbl 112hp—manual trans
FM—235ci I-6 1bbl 112hp—auto trans
TM—235ci I-6 1bbl 112hp—auto trans

3600 Series
FXA—235ci I-6 1bbl 112hp—manual trans
TXA—235ci I-6 1bbl 112hp—manual trans
FUA—235ci I-6 1bbl 112hp—manual trans
TUA—235ci I-6 1bbl 112hp—manual trans
FMA—235ci I-6 1bbl 112hp—auto trans
TMA—235ci I-6 1bbl 112hp—auto trans

3800 Series
FXB—235ci I-6 1bbl 112hp—manual trans
TXB—235ci I-6 1bbl 112hp—manual trans
FUB—235ci I-6 1bbl 112hp—manual trans
TUB—235ci I-6 1bbl 112hp—manual trans
FMB—235ci I-6 1bbl 112hp—auto trans
TMB—235ci I-6 1bbl 112hp—auto trans

Transmission Codes

Code	Type	Plant
M	3 speed	Muncie
S	3 speed	Saginaw
W	3 speed HD	Warner Gear
T	4 speed	Toledo
C	Powerglide	Cleveland

Axle Identification

Code	Ratio	Plant
1500 Series		
AA	3.70:1	G&A
BA	3.70:1	Buffalo
AC	3.55:1	G&A
BC	3.55:1	Buffalo
AB	4.11:1	G&A
BB	4.11:1	Buffalo
3100 Series		
NU	3.90:1	G&A
NV	3.90:1	Buffalo
NA	3.90:1	G&A
NB	3.90:1	Buffalo
3600 Series		
G	4.57:1	G&A
H	4.57:1	Buffalo
D	5.14:1	G&A
C	5.14:1	G&A
Q	5.14:1	G&A
3800 Series		
J	5.14:1	G&A
Q	5.14:1	G&A

Exterior Color Codes

Juniper Green	STD
Onyx Black (Sedan Delivery)	STD
India Ivory	231A
Shadow Gray	231B*
Neptune Green	231C*
Sea Mist Green	231D
Glacier Blue	231E
Skyline Blue	231F*
Gypsy Red	231H*
Commercial Red	234A
Jet Black	234D
Omaha Orange	234E
Mariner Blue	234G
Cream Medium	234L

Yukon Yellow	234M
Ocean Green	234N
Transport Blue	234P
Coppertone	234R
Autumn Brown	234S
Pure White	234T
Bombay Ivory	483-439
*Special Order	

Regular Production Options

3100 Series

3102 Chassis w/ flat-face cowl	$1,087
3103 Chassis and cab	1,357
3104 Pickup	1,430
3105 Panel	1,642
3106 Carryall suburban w/panel type rear doors	1,968
3107 Canopy express	1,699
3112 Chassis w/cowl and windshield	1,120
3116 Carryall suburban w/end-gate	1,968

3600 Series

3602 Chassis w/flat-face cowl	1,238
3603 Chassis and cab	1,497
3604 Pickup	1,593
3608 Chassis and cab w/platform body	1,598
3809 Chassis and cab w/stake body	1,665
3612 Chassis w/cowl and windshield	1,260

3800 Series

3802 Chassis w/flat-face cowl	1,336
3803 Chassis and cab	1,593
3804 Pickup	1,716
3805 Panel	1,940
3807 Canopy express	2,023
3808 Platform (stake pocket rear)	1,733
3809 Chassis and cab w/stake body	2,005
3812 Chassis w/cowl and windshield	1,359

Option Number

100	Directional signals
101	Heater
200	Double-acting shock absorbers
207	Long running boards and rear fenders
208	Rear axle ratio 5.14:1
210	Rear-view mirrors and brackets, LH/RH
211	Rear shock absorber shields
214	Propeller shaft brake
216	Oil-bath air cleaner
217	Positive crankcase ventilation
218	Rear bumper
227	Heavy-duty clutch
230	High sill body
237	Oil filter
241	Governor
249	Dual tail and stop lights
254	Heavy duty rear springs
256	Heavy duty radiator
263	Auxiliary seat
264	Ride control seat
267	Auxiliary rear springs
281	Vacuum Reserve Tank
314	Hydramatic 4-speed automatic transmission
315	Overdrive transmission
316	HD 3-speed transmission
318	HD 4-speed transmission
320	Constant speed electric windshield wiper
324	Power steering
325,326	High output generator
327	Solenoid starter
329	Junior school bus equipment
340-341	Side mounted wheel carrier
345	Heavy duty battery
387	Rear corner windows
390	Deluxe equipment
393	Chrome equipment
395	Left hand door and side mounted spare wheel key lock
398	E-Z Eye glass (tinted shaded)
399	E-Z Eye glass (deluxe cabs)
412	Power brakes
417	Positive crankcase ventilation
423	Running boards
430,431	Custom cabs/custom panels
438,439	Two-tone cab paint colors

Facts

The 1954 trucks were carried over for part of the 1955 model year, as the new, redesigned 1955 trucks weren't quite ready. These early versions are known as the first series. Essentially the same truck as the 1954 model, the it came with a Bombay Ivory painted grille with Onyx Black Chevrolet lettering. The bow-tie emblem was striped in white rather than red. The hoodside emblems were a different design, with Chevrolet lettering beneath the numerical designation. On two-tone trucks, the upper-body color was Bombay Ivory.

Mechanically, the biggest change was the use of an open Hotchkiss drive shaft rather than the previous enclosed torque tube type.

On March 25, 1955, the Advanced Design era ended as the new Task Force trucks were introduced.

The sedan delivery shared the new, redesigned 1955 Chevrolet passenger-car body. Besides a 235.5ci six-cylinder engine, it could be optioned with Chevrolet's new small-block V-8. (For more details on the 1955 sedan delivery, consult the Tri-Chevy Red Book, which covers 1955—57 passenger-car Chevrolets.)

Chapter 11

(Second Series) 1955 Trucks

Production

Sedan Delivery	8,811
3124 Cameo pickup	5,520
Calendar year	Total
1955	393,312

Serial numbers

Description

D255A000001

D—Series: D—1500, H—3100, M—3200, J—3600, L—3800

2—Second Series

55—Year, 55—1955

A—Assembly Plant Code: A—Atlanta GA, B—Baltimore MD, F—Flint MI, J—Janesville WI, K—Kansas City MO, L—Los Angeles CA, N—Norwood OH, O—Oakland CA, S—St. Louis MO, T—Tarrytown NY

000001—Consecutive Sequence Number

Model, Wheelbase & GVW

Model Number and Description	Wheelbase (in)	GVW
1500 Series		
1508 Sedan Delivery	115	4,100
3100 Series		
3102 Chassis w/flat-face cowl	114	5,000
3103 Chassis and cab	114	5,000
3104 Pickup	114	5,000
3105 Panel	114	5,000
3106 Carryall suburban w/panel type rear doors	114	5,000
3112 Chassis w/cowl and windshield	114	5,000
3116 Carryall suburban w/end-gate	114	5,000
3124 Suburbanite (Cameo)	114	5,000
3200 Series		
3204 Pickup	123¼	5,000
3600 Series		
3602 Chassis w/flat-face cowl	123¼	6,900
3603 Chassis and cab	123¼	6,900
3604 Pickup	123¼	6,900

3608 Chassis and cab w/platform body	123$\frac{1}{4}$	6,900
3609 Chassis and cab w/stake body	123$\frac{1}{4}$	6,900
3612 Chassis w/cowl and windshield	123$\frac{1}{4}$	6,900

3800 Series

3802 Chassis w/flat-face cowl	135	8,800
3803 Chassis and cab	135	8,800
3804 Pickup	135	8,800
3805 Panel	135	7,000
3808 Platform	135	8,800
3809 Chassis and cab w/stake body	135	8,800
3812 Chassis w/cowl and windshield	135	8,800

Engine & Transmission Suffix Codes

1500 Series

FZ—235ci I-6 1bbl 123hp—manual trans
TZ—235ci I-6 1bbl 123hp—manual trans
FZC—235ci I-6 1bbl 123hp—manual trans
TZC—235ci I-6 1bbl 123hp—manual trans
FZE—235ci I-6 1bbl 123hp—manual trans
TZE—235ci I-6 1bbl 123hp—manual trans
FZH—235ci I-6 1bbl 123hp—auto trans
TZH—235ci I-6 1bbl 123hp—auto trans
FZJ—235ci I-6 1bbl 123hp—auto trans
TZJ—235ci I-6 1bbl 123hp—auto trans
FYE—235ci I-6 1bbl 123hp—auto trans
TYE—235ci I-6 1bbl 123hp—auto trans
FG—265ci V-8 2bbl 162hp—manual trans
TG—265ci V-8 2bbl 162hp—manual trans
FGC—265ci V-8 2bbl 162hp—O/D manual trans
TGC—265ci V-8 2bbl 162hp—O/D manual trans
FGD—265ci V-8 4bbl 180hp—manual trans
TGD—265ci V-8 4bbl 180hp—manual trans
FGE—265ci V-8 4bbl 180hp—O/D manual trans
TGE—265ci V-8 4bbl 180hp—O/D manual trans
FGF—265ci V-8 2bbl 162hp—manual trans
TGF—265ci V-8 2bbl 162hp—manual trans
FGG—265ci V-8 4bbl 180hp—manual trans
TGG—265ci V-8 4bbl 180hp—manual trans
FGJ—265ci V-8 2bbl 162hp—manual trans
TGJ—265ci V-8 2bbl 162hp—manual trans
FGK—265ci V-8 2bbl 162hp—manual trans
TGK—265ci V-8 2bbl 162hp—manual trans
FGL—265ci V-8 4bbl 180hp—manual trans
TGL—265ci V-8 4bbl 180hp—manual trans
FGM—265ci V-8 4bbl 180hp—manual trans
TGM—265ci V-8 4bbl 180hp—manual trans
FF—265ci V-8 2bbl 162hp—auto trans
TF—265ci V-8 2bbl 162hp—auto trans
FFB—265ci V-8 4bbl 180hp—auto trans
TFB—265ci V-8 4bbl 180hp—auto trans
FFC—265ci V-8 2bbl 162hp—auto trans
TFC—265ci V-8 2bbl 162hp—auto trans
FFD—265ci V-8 4bbl 180hp—auto trans
TFD—265ci V-8 4bbl 180hp—auto trans

3100/3200 Series

FX—235ci I-6 1bbl 123hp—manual trans
TX—235ci I-6 1bbl 123hp—manual trans
FSC—235ci I-6 1bbl 123hp—manual trans
TSC—235ci I-6 1bbl 123hp—manual trans
FM—235ci I-6 1bbl 123hp—auto trans
TM—235ci I-6 1bbl 123hp—auto trans

3600 Series

FXA—235ci I-6 1bbl 123hp—manual trans
TXA—235ci I-6 1bbl 123hp—manual trans
FSA—235ci I-6 1bbl 123hp—manual trans
TSA—235ci I-6 1bbl 123hp—manual trans
FMA—235ci I-6 1bbl 123hp—auto trans
TMA—235ci I-6 1bbl 123hp—auto trans
FE—265ci V-8 2bbl 145hp—manual trans
TE—265ci V-8 2bbl 145hp—manual trans
FEA—265ci V-8 2bbl 145hp—manual trans
TEA—265ci V-8 2bbl 145hp—manual trans

3800 Series

FXB—235ci I-6 1bbl 123hp—manual trans
TXB—235ci I-6 1bbl 123hp—manual trans
FSB—235ci I-6 1bbl 123hp—manual trans
TSB—235ci I-6 1bbl 123hp—manual trans
FMB—235ci I-6 1bbl 123hp—auto trans
TMB—235ci I-6 1bbl 123hp—auto trans
FEB—265ci V-8 2bbl 145hp—manual trans
TEB—265ci V-8 2bbl 145hp—manual trans

Transmission Codes

Code	Type	Plant
M	3 speed/OD	Muncie
S	3 speed/OD	Saginaw
W	3 speed HD	Warner Gear
T	4 speed	Toledo
C	Powerglide	Cleveland

Axle Identification

Code	Ratio
1500 Series	
AA/BA	3.70:1
AC/BC	4.11:1
AB/BB	3.55:1
3100 Series	
AF	3.90:1
AG	4.11:1
3600 Series	
CC	4.57:1
CD	4.57:1

3800 Series

CF	5.14:1
CB	5.14:1

Exterior Color Codes

Juniper Green	STD
Onyx Black (Sedan Delivery)	STD
India Ivory	231A
Shadow Gray	231B*
Neptune Green	231C*
Sea Mist Green	231D
Glacier Blue	231E
Gypsy Red	231H*
Cashmere Blue	231J
Commercial Red	234A
Jet Black	234D
Omaha Orange	234E
Mariner Blue	234G
Cream Medium	234L
Yukon Yellow	234M
Ocean Green	234N
Transport Blue	234P
Coppertone	234R
Autumn Brown	234S
Pure White	234T
Bombay Ivory	443-446
*Special Order	

Regular Production Options

1508 Sedan Delivery $1,699

3100 Series

3102 Chassis w/ flat-face cowl	1,156
3103 Chassis and cab	1,423
3104 Pickup	1,519
3105 Panel	1,801
3106 Carryall suburban w/panel type rear doors	2,150
3112 Chassis w/cowl and windshield	1,193
3116 Carryall suburban w/end-gate	2,150
3124 Suburbanite (Cameo)	1,981

3200 Series

3204 Pickup	1,540

3600 Series

3602 Chassis w/flat-face cowl	1,316
3603 Chassis and cab	1,583
3604 Pickup	1,690
3608 Chassis and cab w/platform body	1,711
3609 Chassis and cab w/stake body	1,708
3612 Chassis w/cowl and windshield	1,353

3800 Series

3802 Chassis w/flat-face cowl	1,444
3803 Chassis and cab	1,711
3804 Pickup	1,844
3805 Panel	2,135
3808 Platform	1,859
3809 Chassis and cab w/stake body	1,944
3812 Chassis w/cowl and windshield	1,481

Option Number

100	Directional signals
101	Heater
200	Double-acting shock absorbers
210	Rear-view mirrors and brackets, LH/RH
211	Rear shock absorber shields
213	Hydrovac power brake
214	Propeller shaft brake
216	Oil-bath air cleaner
217	Positive crankcase ventilation
218	Rear bumper
221	Turbo-Fire V-8 engine (manual trans)
222	Turbo-Fire V-8 engine (O/D trans)
223	Turbo-Fire V-8 engine (PG)
227	Heavy-duty clutch
230	High body sills
237	Oil filter
241	Governor
249	Dual tail and stop lights
254	Heavy duty rear springs
256	Heavy duty radiator
258	Foam rubber seat cushion
263	Auxiliary seat
267	Auxiliary rear springs
269	Airmatic seat
291	Wide base wheels
313	Powerglide automatic transmission
314	Hydramatic 4-speed automatic transmission
315	Overdrive transmission
316	HD 3-speed transmission
318	HD 4-speed transmission
320	Electric windshield wiper
324	Power steering
325,326	High output generator
341	Side mounted wheel carrier
345	Heavy duty battery
350	Power steering
367	Front bumper
384	Spare wheel/carrier
393	Chrome equipment
394	Full view rear window
395	Left hand door and side mounted spare wheel key lock
398	E-Z Eye glass (tinted shaded)
399	E-Z Eye glass (deluxe cabs)
408	Trademaster V-8 engine, 3100-3600 Series
409	Trademaster V-8 engine, 3800 Series
411	Four-barrel carburetor
412	Hydrovac
417	Positive crankcase ventilation
423	Running boards
430,431	Custom cabs/custom panels
450	Air conditioning

Facts

The 1955—59 models are known as the Task Force trucks. These were totally redesigned and updated, including the frame, which used parallel beams with six cross-members. Like the Chevrolet passenger cars, the trucks now used a curved wraparound windshield. Gone were the running boards—at least the part that used to be visible under the doors; a small section, or step, was retained between the rear of the cab and the rear fender. This small section or step remnant would later be used to refer to this type of pickup as a Stepside.

Standard trucks came with the front grille, bumper, and headlights bezels all painted Bombay Ivory. painted front grille, bumper, and headlight bezels. On the Deluxe and Cameo models, these items were chrome.

The inside of the cab was also redesigned. A speedometer with an odometer, along with an ammeter and oil, temperature, and fuel gauges, was located in a V-shaped area behind the steering wheel on the dash panel. An optional radio would be located in the center of the dash with an ashtray on either side.

The standard seats were finished in black with beige trim. The inside of the doors was painted beige. Black waffle-pattern vinyl was used as a headliner.

Deluxe-equipped trucks got a beige/brown seat and dash, and the steering wheel, steering column, and windshield garnish moldings painted brown. The door trim panels were painted brown as well. Also included were chrome knobs, dual sun visors, dual armrests, and a cigar lighter.

On the exterior, brightmetal moldings were used around the windshield, the vent windows, the side door window frames, and the window areas. Also included were dual horns, horns and a chrome grille, headlight bezels, and hubcaps. All Deluxe cabs were also equipped with a larger Panoramic rear window.

Two new models were introduced. Model 3204 was a 1/2-ton pickup that used the longer wheelbase and box of the 3/4-ton trucks. Model 3124 was the Cameo Carrier. Easily recognizable from other 1955 pickups, the Cameo was equipped with fiberglass panels on the sides of the box and the tailgate, side box panels and tailgate panel to give the truck a totally different look, like that of the Fleetside cabs that would come several years later.

The Cameo rear bumper and taillights were also unique. The rear bumper concealed a hidden compartment that contained the spare tire. The rear tailgate also had a Chevrolet bow-tie emblem.

All 1955 Cameos were painted Bombay Ivory with Commercial Red on the rear cab posts. The interior was finished in red-and-white cloth that was similar to the upholstery used on the 1953 Bel Air.

Two-tone paint was available on both standard and Deluxe models. On standard models, the contrasting color was applied on the roof, whereas on Deluxe models, it was applied on the window area. Bombay Ivory was the contrasting color except with Russet Brown, which came with Sand Beige as the contrasting color.

All the trucks that came with a single-color paint scheme had black painted wheels, whereas two-tones had colored wheels. The 1/2-ton two-tone trucks also had stripes on the wheels.

The standard engine was the 235ci six-cylinder rated at 123hp. Optional was the new Chevy small-block V-8 displacing 265ci, rated at 145hp.

Deleted were the Canopy Express models, owing to poor demand.

New options were power steering and power brakes.

All 1955 trucks came with a 12-volt electrical system.

Chapter 12

1956 Trucks

Production

1508 Sedan Delivery	9,445
3124 Cameo Carrier	1,452
Calendar year	Total
1956	353,509

Serial numbers

Description

3A56A000001

3A*—Series: D—1500, 3A—3100, 3B—3200, 3E—3600, 3G—3800

56—Year, 56—1956

A—Assembly Plant Code: A—Atlanta GA, B—Baltimore MD, F—Flint MI, J—Janesville WI, K—Kansas City MO, L—Los Angeles CA, N—Norwood OH, O—Oakland CA, S—St. Louis MO, T—Tarrytown NY

000001—Consecutive Sequence Number

*a "V" prefix indicates a V-8 engine

Model, Wheelbase & GVW

Model Number and Description	Wheelbase (in)	GVW
1500 Series		
1508 Sedan Delivery	115	4,100
3100 Series		
3102 Chassis w/flat-face cowl	114	5,000
3103 Chassis and cab	114	5,000
3104 Pickup	114	5,000
3105 Panel	114	5,000
3106 Carryall suburban w/panel type rear doors	114	5,000
3112 Chassis w/cowl and windshield	114	5,000
3116 Carryall suburban w/end-gate	114	5,000
3124 Suburbanite (Cameo)	114	5,000
3200 Series		
3204 Pickup	123¼	5,000
3600 Series		
3602 Chassis w/flat-face cowl	123¼	6,900
3603 Chassis and cab	123¼	6,900
3604 Pickup	123¼	6,900

3608	Chassis and cab w/platform body	123¼	6,900
3609	Chassis and cab w/stake body	123¼	6,900
3612	Chassis w/cowl and windshield	123¼	6,900

3800 Series

3802	Chassis w/flat-face cowl	135	8,800
3803	Chassis and cab	135	8,800
3804	Pickup	135	8,800
3805	Panel	135	7,000
3808	Platform	135	8,800
3809	Chassis and cab w/stake body	135	8,800
3812	Chassis w/cowl and windshield	135	8,800

Engine & Transmission Suffix Codes

1500 Series

FZ—235ci I-6 1bbl 140hp—manual trans
TZ—235ci I-6 1bbl 140hp—manual trans
FZC—235ci I-6 1bbl 140hp—manual trans
TZC—235ci I-6 1bbl 140hp—manual trans
FZE—235ci I-6 1bbl 140hp—manual trans
TZE—235ci I-6 1bbl 140hp—manual trans
FY—235ci I-6 1bbl 140hp—auto trans
TY—235ci I-6 1bbl 140hp—auto trans
FYE—235ci I-6 1bbl 140hp—auto trans
TYE—235ci I-6 1bbl 140hp—auto trans
FG—265ci V-8 2bbl 162hp—manual trans
TG—265ci V-8 2bbl 162hp—manual trans
FGC—265ci V-8 2bbl 162hp—manual trans
TGC—265ci V-8 2bbl 162hp—manual trans
FGQ—265ci V-8 4bbl 205hp—O/D manual trans
TGQ—265ci V-8 4bbl 205hp—O/D manual trans
FGE—265ci V-8 4bbl 205hp—O/D manual trans
TGE—265ci V-8 4bbl 205hp—O/D manual trans
FGF—265ci V-8 2bbl 162hp—manual trans
TGF—265ci V-8 2bbl 162hp—manual trans
FGJ—265ci V-8 4bbl 180hp—manual trans
TGJ—265ci V-8 4bbl 180hp—manual trans
FGK—265ci V-8 2bbl 162hp—manual trans
TGK—265ci V-8 2bbl 162hp—manual trans
FGL—265ci V-8 4bbl 205hp—manual trans
TGL—265ci V-8 4bbl 205hp—manual trans
FGM—265ci V-8 4bbl 205hp—manual trans
TGM—265ci V-8 4bbl 205hp—manual trans
FGN—265ci V-8 4bbl 205hp—manual trans
TGN—265ci V-8 4bbl 205hp—manual trans
FF—265ci V-8 2bbl 162hp—auto trans
TF—265ci V-8 2bbl 162hp—auto trans
FFB—265ci V-8 4bbl 205hp—auto trans
TFB—265ci V-8 4bbl 205hp—auto trans
FFC—265ci V-8 2bbl 162hp—auto trans
TFC—265ci V-8 2bbl 162hp—auto trans
FFD—265ci V-8 4bbl 205hp—auto trans
TFD—265ci V-8 4bbl 205hp—auto trans

3100/3200 Series

FX—235ci I-6 1bbl 140hp—manual trans
TX—235ci I-6 1bbl 140hp—manual trans
FVC—235ci I-6 1bbl 140hp—manual trans
TVC—235ci I-6 1bbl 140hp—manual trans
FXG—235ci I-6 1bbl 140hp—auto trans
TXG—235ci I-6 1bbl 140hp—auto trans
FA—265ci V-8 2bbl 155hp—manual trans
TA—265ci V-8 2bbl 155hp—manual trans
FB—265ci V-8 2bbl 155hp—auto trans
TB—265ci V-8 2bbl 155hp—auto trans
FM—265ci V-8 2bbl 155hp—manual trans
TM—265ci V-8 2bbl 155hp—manual trans

3600 Series

FXA—235ci I-6 1bbl 140hp—manual trans
TXA—235ci I-6 1bbl 140hp—manual trans
FVA—235ci I-6 1bbl 140hp—manual trans
TVA—235ci I-6 1bbl 140hp—manual trans
FXGA—235ci I-6 1bbl 140hp—manual trans
TXGA—235ci I-6 1bbl 140hp—manual trans
FAA—265ci V-8 2bbl 155hp—manual trans
TAA—265ci V-8 2bbl 155hp—manual trans
FBA—265ci V-8 2bbl 155hp—auto trans
TBA—265ci V-8 2bbl 155hp—auto trans
FMA—265ci V-8 2bbl 155hp—manual trans
TMA—265ci V-8 2bbl 155hp—manual trans

3800 Series

FXB—235ci I-6 1bbl 140hp—manual trans
TXB—235ci I-6 1bbl 140hp—manual trans
FVB—235ci I-6 1bbl 140hp—manual trans
TVB—235ci I-6 1bbl 140hp—manual trans
FXGB—235ci I-6 1bbl 140hp—auto trans
TXGB—235ci I-6 1bbl 140hp—auto trans
FAB—265ci V-8 2bbl 155hp—manual trans
TAB—265ci V-8 2bbl 155hp—manual trans
FBB—265ci V-8 2bbl 155hp—auto trans
TBB—265ci V-8 2bbl 155hp—auto trans
FMB—265ci V-8 2bbl 155hp—manual trans

Transmission Codes

Code	Type	Plant
M	3 speed	Muncie
S	3 speed	Saginaw
W	3 speed HD	Warner Gear
T	4 speed	Toledo
C	Powerglide	Cleveland

3100/3200 Series
CH56HK   Hydra-matic

3600 Series
CH56   Hydra-matic

3800 Series
CHC56   Hydra-matic

Axle Identification

Code	Ratio

1500 Series
AA/BA	3.70:1
AC/BC	4.11:1
AB/BB	3.55:1

3100/3200 Series
AF	3.90:1
AG	4.11:1

3600 Series

CC	4.57:1
CD	4.57:1

3800 Series

CF	5.14:1
CB	5.14:1

Exterior Color Codes

Cardinal Red	234A
Regal Blue	234B
Sand Beige	234C
Jet Black	234D
Omaha Orange	234E
Granite Gray	234F
Empire Blue	234G
Golden Yellow	234L
Yukon Yellow	234M
Ocean Green	234N
Crystal Blue	234P
Pure White (w/chrome trim)	234T
Pure White (w/std trim)	234U
Pure White (w/std trim, 3600/3800)	234V
India Ivory	500A*
Pinecrest Green	504A*
Sherwood Green	506A*
Nassau Blue	508A*
Harbor Blue	510A*
Matador Red	522A*

*Model 1508 w/std trim

Regular Production Options

1508 Sedan Delivery	$1,865

3100 Series

3102 Chassis w/ flat-face cowl	1,303
3103 Chassis and cab	1,567
3104 Pickup	1,670
3105 Panel	1,966
3106 Suburban carryall w/panel rear doors	2,300
3112 Chassis w/cowl and windshield	1,341
3116 Suburban carryall w/end-gate	2,300
3124 Suburbanite (Cameo)	2,144

3200 Series

3204 Pickup	1,692

3600 Series

3602 Chassis w/flat-face cowl	1,481
3603 Chassis and cab	1,745
3604 Pickup	1,858
3809 Chassis and cab w/stake body	1,950
3612 Chassis w/cowl and windshield	1,519

3800 Series

3802 Chassis w/flat-face cowl	1,611
3803 Chassis and cab	1,875
3804 Pickup	2,009
3805 Panel	2,327
3809 Chassis and cab w/stake body	2,122
3812 Chassis w/cowl and windshield	1,649

Option Number

101	Heater
105,106	Directional signals
110,111	Air conditioning
200	Double-acting shock absorbers
210	Rear-view mirrors and brackets, LH/RH
211	Rear shock absorber shields
213	Hydrovac power brake
216	Oil-bath air cleaner
217	Positive crankcase ventilation
218	Rear bumper
227	Heavy-duty clutch
230	High body sills
237	Oil filter
241	Governor
249	Dual tail and stop lights
254	Heavy duty rear springs
256	Heavy duty radiator
258	Foam rubber seat cushion
263	Auxiliary seat
267	Auxiliary rear springs
306,307	Speedometer drive fittings
314	Hydra-matic 4-speed automatic transmission
315	Overdrive transmission
316	HD 3-speed transmission
318	HD 4-speed transmission
320	Electric windshield wiper
324	Power steering
325,326	High output generator
341	Side mounted wheel carrier
345	Heavy duty battery
350,351	Power steering
367	Front bumper
384	Spare wheel/carrier

The 1956 Advance Design Series pickup.

Facts

Changes were minimal for 1956. The hood emblem was slightly different, with its side extensions stretching ou from the bottom rather than the top. V-8—equipped trucks came with a hood emblem that incorporated a V below the bow-tie emblem.

Horsepower was up to 140 on the 235ci six-cylinder and 155 on the 265ci Trademaster V-8.

The Cameo Carrier pickup could be had in eight two-tone color combinations: Cardinal Red/Bombay Ivory, Cardinal Red/Sand Beige, Jet Black/Golden Yellow, Arabian Ivory/Cardinal Red, Arabian Ivory/Regal Blue, Arabian Ivory/Granite Gray, Arabian Ivory/Ocean Green, and Arabian and Arabian Ivory/Crystal Blue. Production was only 1,452 units.

Chapter 13

1957 Trucks

Production

1508 Sedan Delivery	7,273
3124 Cameo Carrier	2,244
Calendar year	Total
1957	351,739

Serial numbers

Description

3A57A000001

3A*—Series: D—1508, 3A—3100, 3B—3200, 3E—3600, 3G—3800

57—Year, 57—1957

A—Assembly Plant Code: A—Atlanta GA, B—Baltimore MD, F—Flint MI, J—Janesville WI, K—Kansas City MO, L—Los Angeles CA, N—Norwood OH, O—Oakland CA, S—St. Louis MO, T—Tarrytown NY

000001—Consecutive Sequence Number

*a "V" prefix indicates a V-8 engine

Model, Wheelbase & GVW

Model Number and Description	Wheelbase (in)	GVW
1500 Series		
1508 Sedan Delivery 6 & 8cyl	115	4,100
3100 Series		
3102 Chassis w/flat-face cowl	114	5,000
3103 Chassis and cab	114	5,000
3104 Pickup	114	5,000
3105 Panel	114	5,000
3106 Suburban carryall w/panel type rear doors	114	5,000
3112 Chassis w/cowl and windshield	114	5,000
3116 Suburban carryall w/end-gate	114	5,000
3124 Suburbanite p/u (Cameo)	114	5,000
3200 Series		
3204 Pickup	123¼	5,000
3600 Series		
3602 Chassis w/flat-face cowl	123¼	6,900
3603 Chassis and cab	123¼	6,900
3604 Pickup	123¼	6,900

| 3809 Chassis and cab w/stake body | 123¼ | 6,900 |
| 3612 Chassis w/cowl and windshield | 123¼ | 6,900 |

3800 Series

3802 Chassis w/flat-face cowl	135	9,600
3803 Chassis and cab	135	9,600
3804 Pickup	135	7,000
3805 Panel	135	7,000
3809 Chassis and cab w/stake body	135	9,600
3812 Chassis w/cowl and windshield	135	9,600

Engine & Transmission Suffix Codes

1508 Series

FA—235ci I-6 1bbl 140hp—manual trans
TA—235ci I-6 1bbl 140hp—manual trans
FBC—235ci I-6 1bbl 140hp—auto trans
TBC—235ci I-6 1bbl 140hp—auto trans
FAD—235ci I-6 1bbl 140hp—manual trans
TAD—235ci I-6 1bbl 140hp—manual trans
FC—235ci V-8 1bbl 162hp—manual trans
TC—235ci V-8 1bbl 162hp—manual trans
FCD—235ci V-8 1bbl 162hp—manual trans
TCD—235ci V-8 1bbl 162hp—manual trans
FCE—265ci V-8 2bbl 162hp—auto trans
TCE—265ci V-8 2bbl 162hp—auto trans
FE—283ci V-8 2bbl 220hp—manual trans
TE—265ci V-8 2bbl 220hp—manual trans
FEA—265ci V-8 4bbl 245hp—manual trans*
TEA—283ci V-8 4bbl 245hp—manual trans*
FEB—283ci V-8 4bbl 270hp—manual trans*
TEB—283ci V-8 2bbl 270hp—manual trans*
FEC—283ci V-8 2bbl 220hp—O/D manual trans
TEC—283ci V-8 4bbl 220hp—O/D manual trans
FF—283ci V-8 2bbl 185hp—Powerglide
TF—283ci V-8 2bbl 185hp—Powerglide
FFC—283ci V-8 2bbl 220hp—Powerglide
TFC—283ci V-8 4bbl 220hp—Powerglide
FFD—283ci V-8 4bbl 245hp—Powerglide*
TFD—283ci V-8 4bbl 245hp—Powerglide*
FFA—283ci V-8 2bbl 185hp—Powerglide
TFA—283ci V-8 2bbl 185hp—Powerglide
FFE—283ci V-8 4bbl 220hp—Powerglide
TFE—283ci V-8 2bbl 220hp—Powerglide
FFJ—283ci V-8 F.I. 250hp—Powerglide
TFJ—283ci V-8 F.I. 250hp—Powerglide
FG—283ci V-8 2bbl 185hp—Turboglide
TG—283ci V-8 2bbl 185hp—Turboglide
FGC—283ci V-8 4bbl 220hp—Turboglide
TGC—283ci V-8 4bbl 220hp—Turboglide
FGD—283ci V-8 4bbl 245hp—Turboglide
TGD—283ci V-8 4bbl 245hp—Turboglide
FGF—283ci V-8 F.I. 250hp—Turboglide
TGF—283ci V-8 F.I. 250hp—Turboglide
FEJ—283ci V-8 F.I. 250hp—manual
TEJ—283ci V-8 F.I. 250hp—manual

FEK—283ci V-8 F.I. 283hp—manual
TEK—283ci V-8 F.I. 283hp—manual
*Dual four barrel

3100/3200/3600/3800 Series
FH—235ci I-6 1bbl 140hp—manual trans
TH—235ci I-6 1bbl 140hp—manual trans
FHE—235ci I-6 1bbl 140hp—manual trans
FHD—235ci I-6 1bbl 140hp—manual trans
THD—235ci I-6 1bbl 140hp—HD manual trans
FLG—235ci I-6 1bbl 140hp—HD manual trans
TL—265ci V-8 2bbl 162hp—manual trans
FLA—265ci V-8 2bbl 162hp—manual trans
TLA—265ci V-8 2bbl 162hp—HD manual trans
FLB—265ci V-8 2bbl 162hp—HD manual trans
TLB—265ci V-8 2bbl 162hp—manual trans

Transmission Codes

Code	Type	Plant
M	3 speed	Muncie
S	3 speed O/D	Saginaw
WG	3 speed HD	Warner Gear
T	4 speed	Toledo
T	Turboglide	Toledo
C	Powerglide	Cleveland

Axle Identification

Code	Ratio

1508 Series
AB/BB	3.36:1
AA/BA	3.55:1
AC/BC	4.11:1

3100/3200 Series
AF	3.90:1
AG	4.11:1

3600 Series
CC	4.57:1
CD	4.57:1

3800 Series
CF	5.14:1
CB	5.14:1

Exterior Color Codes

1508 Series
Onyx Black	STD
Imperial Ivory	500A
Surf Green	504A
Highland Green	505A
Larkspur Blue	508A
Harbor Blue	510A
Matador Red	522A

3100/3200/3600/3800 Series
Brester Green	STD
Cardinal Red	234A
Indian Turquoise	234B
Sand Beige	234C
Jet Black	234D
Omaha Orange	234E
Granite Gray	234F
Royal Blue	234G
Golden Yellow	234L
Yukon Yellow	234M
Ocean Green	234N
Alpine Blue	234P
Pure White	234U/V
Sandstone Beige	234W

Regular Production Options
1508 Sedan Delivery	$2,020

3100 Series
3102 Chassis w/ flat-face cowl	1,433
3103 Chassis and cab	1,697
3104 Pickup	1,800
3105 Panel	2,101
3106 Suburban carryall w/panel rear doors	2,435
3112 Chassis w/cowl and windshield	1,471
3116 Suburban carryall w/end-gate	2,435
3124 Cameo Carrier	2,273

3200 Series
3204 Pickup   1,838

3600 Series
3602 Chassis
    w/flat-face cowl   1,616
3603 Chassis and cab   1,880
3604 Pickup   1,993
3809 Chassis and cab
    w/stake body   2,085
3612 Chassis w/cowl
    and windshield   1,654

3800 Series
3802 Chassis
    w/flat-face cowl   1,763
3803 Chassis and cab   2,027
3804 Pickup   2,160
3805 Panel   2,489
3809 Chassis and cab
    w/stake body   2,274
3812 Chassis w/cowl
    and windshield   1,801

Option Number

101	Heater
105	Directional signals
110	Air conditioning (Sedan Delivery)
200	Double-acting shock absorbers
210	Rear-view mirrors and brackets, LH/RH
211	Rear shock absorber shields
213	Hydrovac power brake
216	Oil-bath air cleaner
217	Positive crankcase ventilation
218	Rear bumper
221	Turbo-fire 265ci V-8 engine
223	Turbo-fire 283ci V-8 engine
227	Heavy-duty clutch
230	Platform body
237	Oil filter
241	Governor
253	Heavy duty front springs
254	Heavy duty rear springs
256	Heavy duty radiator
258	Foam rubber seat cushion
263	Auxiliary seat
267	Auxiliary rear springs
306,307	Speedometer drive fittings
313	Powerglide automatic transmission
314	Hydra-matic 4-speed automatic transmission
315	Overdrive transmission
316	HD 3-speed transmission
318	HD 4-speed transmission
320	Electric windshield wiper
324	Power steering
325,326	High output generator
341	Side mounted spare wheel carrier
345	Heavy duty battery
350	Power steering (6cyl)
351	Power steering (V-8)
393	Chrome equipment
394	Full view rear window
395	Left hand door and side mounted spare wheel key lock
398	E-Z Eye glass
408	Trademaster V-8 engine, 3100-3600 Series
409	Trademaster V-8 engine, 3800 Series
410	Four barrel carburetor
412	Power brakes
417	Positive crankcase ventilation
423	Running boards
431	Custom cab
434	Custom panel
482	Full width seat
675,676, 678	Positraction rear axle
690	Four wheel drive

1957 Chevy. Paul G. McLaughlin

Facts

The 1957 models were facelifted with a dual trapezoidal front grille. The hood was flatter and incorporated twin windsplit bulges, and the side fender spears were oval in shape. The front hood medallion was larger, too. Chrome bombsight ornaments were available for the twin hood bulges.

Cameo pickups got a styling embellishment as well. This was a contrasting color that was painted along the side of the pickup box which was and bordered with chrome moldings. At the front of the band was a Chevrolet bow-tie emblem with Cameo script behind it. Production was 2,244 units.

A midyear introduction was the availability of the NAPCO four-wheel-drive transfer case and front driving axle on most 3100, 3600, and 3800 Series models. The NAPCO four-wheel drive was only available on trucks equipped with the 235ci six-cylinder and a four-speed manual.

Chapter 14

1958 Trucks

Production

1171 Sedan Delivery	7,466
3124 Cameo Carrier	1,405
Calendar year	Total
1958	278,615

Serial numbers

Description

3A58A000001

3A*—Series: G—1171, H—1271, 3A—3100, 3B—3200, 3E—3600, 3G—3800

58—Year, 58—1958

A—Assembly Plant Code: A—Atlanta GA, B—Baltimore MD, F—Flint MI, J—Janesville WI, K—Kansas City MO, L—Los Angeles CA, N—Norwood OH, O—Oakland CA, S—St. Louis MO, T—Tarrytown NY, W—Willow Run, MI

000001—Consecutive Sequence Number

*a "V" prefix indicates a V-8 engine

Model, Wheelbase & GVW

Model Number and Description	Wheelbase (in)	GVW
1171/1271 Series		
1171 Sedan Delivery, 6cyl	117½	4,100
1171 Sedan Delivery, 8cyl	117½	4,100
3100 Series		
3102 Chassis & cowl	114	5,000
3103 Chassis and cab	114	5,000
3104 Stepside pickup	114	5,000
3105 Panel delivery	114	5,000
3106 Suburban carryall w/rear doors	114	5,000
3116 Suburban carryall w/end-gate	114	5,000
3124 Cameo Carrier	114	5,000
3134 Fleetside pickup	114	5,000
3153 Chassis and cab, 4WD	114	5,600
3154 Stepside pickup	114	5,600
3155 Panel delivery, 4WD	114	5,600
3184 Fleetside pickup 4WD	114	5,600
3156 Suburban carryall w/rear doors 4WD	114	5,600
3156 Suburban carryall w/rear doors 4WD	114	5,600

3200 Series

3202 Chassis & cowl	123¼	5,000
3203 Chassis & cab	123¼	5,000
3204 Stepside pickup	123¼	5,000
3234 Fleetside pickup	123¼	5,000

3600 Series

3602 Chassis & cowl	123¼	6,900
3603 Chassis and cab	123¼	6,900
3604 Stepside pickup	123¼	6,900
3609 Stake bed	123¼	6,900
3634 Fleetside pickup	123¼	6,900
3653 Chassis & cab 4WD	123¼	7,300
3654 Stepside pickup 4WD	123¼	7,300
3684 Fleetside pickup 4WD	123¼	7,300
3659 Chassis & cab 4WD	123¼	7,300

3800 Series

3802 Chassis & cowl	135	9,600
3803 Chassis and cab	135	9,600
3804 Stepside pickup	135	7,000
3805 Panel delivery	135	7,000
3809 Stake bed	135	9,600
3853 Chassis and cab 4WD	135	7,400
3854 Stepside pickup 4WD	135	7,400
3855 Panel delivery 4WD	135	7,400
3859 Stake bed 4WD	135	7,400

Engine & Transmission Suffix Codes

1171 Series

A—235ci I-6 1bbl 145hp—manual trans
AE—235ci I-6 1bbl 145hp—manual trans
B—235ci I-6 1bbl 145hp—Powerglide

1271 Series

C—283ci V-8 2bbl 185hp—manual trans
CB—283ci V-8 2bbl 185hp—manual trans
CD—283ci V-8 2bbl 185hp—O/D manual trans
CF—283ci V-8 4bbl 230hp—manual trans
CG—283ci V-8 4bbl 230hp—O/D manual trans
CH—283ci V-8 F.I. 250hp—manual trans
D—283ci V-8 2bbl 185hp—Powerglide
DB—283ci V-8 4bbl 230hp—Powerglide
E—283ci V-8 2bbl 185hp—Turboglide
EB—283ci V-8 4bbl 230hp—Turboglide
EC—283ci V-8 F.I. 250hp—Turboglide
F—348ci V-8 4bbl 250hp—manual trans
FA—348ci V-8 3x2bbl 280hp—manual trans
G—348ci V-8 4bbl 250hp—Powerglide
H—348ci V-8 4bbl 250hp—Turboglide
HA—348ci V-8 3x2bbl 280hp—Turboglide

3100/3200/3600/3800 Series

J—235ci I-6 1bbl 145hp—manual trans
JF—235ci I-6 1bbl 145hp—HD manual trans
JC—235ci I-6 1bbl 145hp—manual trans
M—283ci V-8 2bbl 160hp—manual trans
MA—283ci V-8 2bbl 160hp—HD manual trans
MB—283ci V-8 2bbl 160hp—manual trans

Transmission Codes

Code	Type	Plant

1171/1271 Series

S	3 speed O/D	Saginaw
B	Turboglide	Toledo
C	Powerglide	Cleveland

3100/3200/3600/3800 Series

S	3 speed	Saginaw
HS	3 speed O/D	Saginaw
W	4 speed	Warner Gear
CHC55	Hydramatic	—
CHC55	Hydramatic	—

Axle Identification

Code	Ratio

1171/1271 Series

AB, BB, AM	3.36:1
AA, BA, AK	3.55:1
AC, BC, AL	4.11:1

3100/3200 Series

AF	3.90:1
AJ, AR	3.90:1 4WD
AG	4.11:1

3600 Series

CC	4.57:1
CD	4.57:1 dual wheels
CP	4.57:1 4WD

3800 Series

CF	5.14:1
CB	5.14:1 dual wheels
CQ	5.14:1 4WD

Exterior Color Codes

1171/1271 Series

Onyx Black	900
Imperial Ivory	500

Glen Green	903
Forest Green	905
Cashmere Blue	910
Fathom Blue	912
Rio Red	923
Snowcrest White	936
Honey Beige	938

3100/3200/3600/3800 Series

Jet Black	700
Oriental Green	703
Polar Green	704
Glade Green	705
Dawn Blue	707
Marine Blue	708
Tarton Turquoise	710
Kodiak Brown	712
Cardinal Red	714
Omaha Orange	716
Golden Yellow	718
Yukon Yellow	719
Pure White	721
Granite Gray	723

Regular Production Options

1500 Series

1171 Sedan delivery	$2,123

3100 Series

3102 Chassis	1,517
3103 Chassis and cab	1,770
3104 Stepside pickup	1,884
3105 Panel delivery	2,185
3106 Suburban carryall w/rear doors	2,518
3116 Suburban carryall w/end-gate	2,518
3124 Cameo carrier	2,231
3134 Fleetside pickup	1,900

3200 Series

3203 Chassis & cab	1,808
3204 Stepside pickup	1,922
3234 Stepside pickup	1,938

3600 Series

3602	Chassis	1,689
3603	Chassis and cab	1,953
3604	Stepside pickup	2,066
3609	Stake bed	2,518
3634	Fleetside pickup	2,082

3800 Series

3802	Chassis & cowl	1,836
3803	Chassis & cab	2,100
3804	Stepside pickup	2,233
3805	Panel delivery	2,561
3809	Stake bed	2,346

3100/3200/3600/3800 Series

Option Number

112	Heater
115	Heater recirculating
200	Shock absorbers
210	Exterior mirror
211	Rear shock absorber shields
212	Hydrovac power brakes
218	Rear bumper
223	Turbo-fire 283ci V-8 engine
227	Heavy-duty clutch
230	Platform body
237	Oil filter (1 qt.)
241	Governor
253	Heavy duty front springs
254	Heavy duty rear springs (8 leaf)
256	Heavy duty radiator
258	Foam rubber seat cushion
263	Auxiliary seat
267	Heavy duty rear springs (13 leaf)
306,307	Speedometer drive fittings
313	Powerglide automatic transmission
314	Hydra-matic 4-speed automatic transmission
315	Overdrive transmission
316	HD 3-speed transmission
318	4-speed transmission
320	Electric windshield wipers
324	Power steering
325,326	High output generator
341	Side mounted spare wheel carrier
345	Heavy duty battery
350	Power steering (6cyl)
351	Power steering (V-8)

393	Chrome equipment
394	Full view rear window
395	Left hand door and side mounted spare wheel key lock
396	Tire lock
398	E-Z Eye glass
408	Trademaster V-8 engine
423	Running boards
431	Custom panel
587	Generator, heavy duty
591	Air cleaner, oil bath
592	Oil filter (2qt)
599	Right door & tire lock
683	Free wheeling hubs
695	Bostrom seat

1171/1271 Sedan Delivery

103	Radio, manual control
104	Radio, push-button control
110	Air-conditioning
112	Heater
216	Air cleaner
220	Dual exhausts
227	Heavy-duty clutch
237	Oil Filter (1½ qt)
241	Governor
263	Auxiliary seat
302	Turboglide transmission
313	Powerglide transmission
315	Overdrive transmission
320	Electric windshield wipers
324	Power steering
325	Generator (45 amp)
338	Generator (35 amp)
348	Steering wheel (deluxe)
398	Tinted glass (E-Z-Eye)
345	Battery, HD
410	Super Turbo-Fire engine, 230hp
412	Vacuum power brake
417	Positive crankcase ventilation
427	Padded instrument panel
482	Full width seat
576	Turbo-Thrust engine, 250hp
576,573	Super Turbo-Thrust engine, 280hp
576	Special Turbo-Thrust engine, 305hp
576,573	Special Super Turbo-Thrust engine, 315hp
577	Special Turbo-Thrust engine, 320hp
577,574	Special Super Turbo-Thrust engine, 335hp

| 578 | Ramjet Fuel Injection, 250hp | 593 | HD rear coil springs |
| 578 | Special Ramjet Fuel Injection, 290hp | 676 | Positraction rear axle |

A sharp 1958 Chevrolet Apache Pickup was a no sale at a $5,500.00 bid. Paul G. McLaughlin

Facts

For 1958, Chevrolet trucks used a different classification system, whereby the last two numerals of the old designation were dropped. Thus, a 3100 Series truck became a 31 and was so designated on the new fenderside emblems. In addition, all light-duty trucks up to 9,600lb GVW were now called Apache. Medium-duty trucks with a maximum GVW of 21,000lb were called Viking, and heavy-duty trucks up to 25,000lb GVW were called Spartan.

The trucks got a new front end look. This consisted of a quad headlight system, and a new grille, hood, and fenders. The C-h-e-v-r-o-l-e-t letters were stamped on the horizontal center grille bar. bar in the center of the grille. As before, Deluxe-optioned trucks got a chrome grille.

Although the Cameo pickup was briefly available, the biggest news was the introduction of the Fleetside pickups. Whereas the Cameo was a disguised Stepside, the Fleetside was Chevrolet's first slab-sided wide-box pickup. The box was made of steel and featured a double-wall contruction. A Fleetside nameplate was used on the rear upper corners of the box. Fleetsides were available in 1/2- and 3/4-ton models with 6-1/2ft or 8ft boxes.

The standard 235ci six-cylinder got a compression boost to 8.25:1 resulting in 145hp. The 265ci V-8 was replaced by a larger, 160hp, version displacing 283ci.

Air conditioning became optional from the factory. Previously, it was a dealer-installed option.

The three-speed overdrive manual transmission was deleted from the option list.

Bob Inskeep of Farmington, New Mexico, owns one of the last cameo carriers built by Chevrolet. This 1958 model is one of about 400 built that year. Paul G. McLaughlin

Chapter 15

1959 Trucks

Production

1170,1270 2dr Sedan Delivery	5,266
1180,1280 2dr Sedan Pickup	
(El Camino)	23,837
Calendar year	Total
1959	326,093

Serial numbers

Description

3A58A000001

3A*—Series: G—1100, H—1200, 3A—3100, 3B—3200, 3E—3600, 3G—3800

59—Year, 59—1959

A—Assembly Plant Code: A—Atlanta GA, B—Baltimore MD, F—Flint MI, J—Janesville WI, K—Kansas City MO, L—Los Angeles CA, N—Norwood OH, O—Oakland CA, S—St. Louis MO, T—Tarrytown NY, W—Willow Run MI

000001—Consecutive Sequence Number

*a "V" prefix indicates a V-8 engine

Location: On plate attached to left door hinge post.

Model, Wheelbase & GVW

Model Number and Description	Wheelbase (in)	GVW
1100/1200 Series		
1170 Sedan Delivery, 6cyl	119	4,900
1270 Sedan Delivery, 8cyl	119	4,900
1180 El Camino 6cyl	119	4,900
1280 El Camino 8cyl	119	4,900
3100 Series		
3102 Chassis & cowl	114	5,000
3103 Chassis and cab	114	5,000
3104 Stepside pickup	114	5,000
3105 Panel delivery	114	5,000
3106 Suburban carryall w/rear doors	114	5,000
3116 Suburban carryall w/end-gate	114	5,000
3134 Fleetside pickup	114	5,000
3153 Chassis and cab, 4WD	114	5,600
3154 Stepside pickup	114	5,600

3155	Panel delivery, 4WD	114	5,600
3184	Fleetside pickup 4WD	114	5,600
3156	Suburban carryall w/rear doors 4WD	114	5,600
3156	Suburban carryall w/rear doors 4WD	114	5,600

3200 Series

3202	Chassis & cowl	123$\frac{1}{4}$	5,000
3203	Chassis & cab	123$\frac{1}{4}$	5,000
3204	Stepside pickup	123$\frac{1}{4}$	5,000
3234	Fleetside pickup	123$\frac{1}{4}$	5,000

3600 Series

3602	Chassis & cowl	123$\frac{1}{4}$	6,900
3603	Chassis & cab	123$\frac{1}{4}$	6,900
3604	Stepside pickup	123$\frac{1}{4}$	6,900
3609	Stake bed	123$\frac{1}{4}$	6,900
3634	Fleetside pickup	123$\frac{1}{4}$	6,900
3653	Chassis & cab 4WD	123$\frac{1}{4}$	7,300
3654	Stepside pickup 4WD	123$\frac{1}{4}$	7,300
3684	Fleetside pickup 4WD	123$\frac{1}{4}$	7,300
3659	Chassis & cab 4WD	123$\frac{1}{4}$	7,300

3800 Series

3802	Chassis & cowl	135	9,600
3803	Chassis & cab	135	9,600
3804	Stepside pickup	135	7,000
3805	Panel delivery	135	7,000
3809	Stake bed	135	9,600
3853	Chassis and cab 4WD	135	7,400
3854	Stepside pickup 4WD	135	7,400
3855	Panel delivery 4WD	135	7,400
3859	Stake bed 4WD	135	7,400

Engine & Transmission Suffix Codes

1100/1200 Series

A—235ci I-6 1bbl 145hp—manual trans
AE—235ci I-6 1bbl 145hp—manual trans
B—235ci I-6 1bbl 145hp—Powerglide
C—235ci V-8 2bbl 185hp—manual trans
CD—283ci V-8 2bbl 185hp—o/d manual trans
CF—283ci V-8 4bbl 230hp—manual trans
CG—283ci V-8 4bbl 230hp—o/d manual trans
CH—283ci V-8 F.I. 250hp—manual trans
CJ—283ci V-8 F.I. 290hp—manual trans
D—283ci V-8 2bbl 185hp—Powerglide
DB—283ci V-8 4bbl 230hp—Powerglide
DK—283ci V-8 2bbl 185hp—Powerglide
DM—283ci V-8 4bbl 230hp—Powerglide
DP—283ci V-8 F.I. 250hp—Powerglide
E—283ci V-8 2bbl 185hp—Turboglide
EB—283ci V-8 4bbl 230hp—Turboglide
EC—283ci V-8 F.I. 250hp—Turboglide
EJ—283ci V-8 4bbl 230hp—Turboglide
F—348ci V-8 4bbl 250hp—manual trans
FA—348ci V-8 3x2bbl 280hp—manual trans
FB—348ci V-8 3x2bbl 315hp—manual trans

FD—348ci V-8 4bbl 300hp—manual trans
FE—348ci V-8 3x2bbl 335hp—manual trans
FG—348ci V-8 4bbl 320hp—manual trans
G—348ci V-8 4bbl 250hp—Powerglide
GB—348ci V-8 3x2bbl 280hp—Powerglide
GD—348ci V-8 4bbl 305hp—Powerglide
H—348ci V-8 4bbl 250hp—Turboglide
HA—348ci V-8 3x2bbl 280hp—Turboglide

3100/3200/3600/3800 Series
J—235ci I-6 1bbl 145hp—manual trans
JF—235ci I-6 1bbl 135hp—HD manual trans
JC—235ci I-6 1bbl 135hp—manual trans
M—283ci V-8 2bbl 160hp—manual trans
MA—283ci V-8 2bbl 160hp—HD manual trans

Transmission Codes

Code	Type	Plant
B	Turboglide	Toledo
C	Powerglide	Cleveland
M	3 speed	Muncie
S	3 speed O/D	Saginaw
W	4 speed	Warner Gear

Axle Identification

Code	Ratio

1100/1200 Series

AW, BW, AX, BX	3.08:1
AB, BB, AM, BM	3.36:1
AA, BA, AK, BK	3.55:1
FE, BE, FH, BH	3.70:1
AC, BC, AL, BL	4.11:1

3100/3200 Series

AY	3.38:1
AJ	3.70:1
AR	3.90:1
AV	4.11:1

3600 Series

CC	4.57:1
CD	4.57:1 dual wheels
CP	4.57:1 4WD
CH	4.57:1

3800 Series

CF	5.14:1
CB	5.14:1 dual wheels
CQ	5.14:1 4WD

Exterior Color Codes

1100/1200 Series

Tuxedo Black	900
Aspen Green	903
Highland Green	905
Frost Blue	910
Harbor Blue	912
Crown Saphire	914
Gothic Gold	920
Roman Red	923
Classic Cream	925
Snowcrest White	936
Satin Beige	938
Grecian Gray	940
Cameo Coral	942

3100/3200/3600/3800 Series

Jet Black	700
Galway Green	703
Sherwood Green	704
Glade Green	705
Dawn Blue	707
Baltic Blue	708
Tarton Turquoise	710
Frontier Beige	712
Cardinal Red	714
Omaha Orange	716
Golden Yellow	718
Yukon Yellow	719
Pure White	721
Cadet Gray	723

Regular Production Options

1170 Sedan Delivery	$2,363
1180 El Camino	2,352

3100 Series

3102 Chassis	1,580
3103 Chassis and cab	1,834
3104 Stepside pickup	1,948
3105 Panel delivery	2,249
3106 Suburban carryall w/rear doors	2,583
3116 Suburban carryall w/end-gate	2,616
3134 Fleetside pickup	1,964

3200 Series

3203 Chassis & cab	1,872
3204 Stepside pickup	1,986
3234 Stepside pickup	2,002

3600 Series

3602 Chassis	1,753
3603 Chassis and cab	2,018
3604 Stepside pickup	2,132
3609 Stake bed	2,223
3634 Fleetside pickup	2,148

3800 Series

3802 Chassis & cowl	1,899
3803 Chassis & cab	2,164
3804 Stepside pickup	2,298
3805 Panel delivery	2,626
3809 Stake bed	2,411

3100/3200/3600/3800 Series

Option Number

103	Radio, manual
112	Heater
115	Heater, recirculating
200	Shock absorbers
210	Exterior mirror
211	Rear shock absorber shields
212	Hydrovac power brakes
218	Rear bumper
227	Heavy-duty clutch
230	Platform body
237	Oil filter (1 qt.)
241	Governor
253	Heavy duty front springs
254	Heavy duty rear springs (8 leaf)
256	Heavy duty radiator
258	Foam rubber seat cushion
263	Auxiliary seat
267	Heavy duty rear springs (13 leaf)
314	Hydra-matic 4-speed automatic transmission
315	Overdrive transmission
316	HD 3-speed transmission
318	4-speed transmission
320	Electric windshield wipers
324	Power steering
336	Generator, 40 amp
338	Generator, 35 amp
341	Side mounted spare wheel carrier
345	Heavy duty battery
350	Power steering (6cyl)
351	Power steering (V-8)
378	Generator, 50 amp
379	Generator, 50 amp, HD
393	Chrome equipment
394	Full view rear window
395	Left hand door and side mounted spare wheel key lock pickup
396	Tire lock
398	E-Z Eye glass
408	Trademaster V-8 engine
423	Running boards
431	Custom panel
587	Generator, heavy duty
591	Air cleaner, oil bath
592	Oil filter (2qt)
599	Right door & tire lock
675	Positraction rear axle
683	Free wheeling hubs
695	Bostrom seat

1100/1200 Sedan Delivery/El Camino

103	Radio, manual control
104	Radio, push-button control
109	Windshield washers
110	Air conditioning
112	Heater
117	Wheel discs
216	Air cleaner, oil bath
220	Dual exhausts
227	Heavy-duty clutch
237	Oil filter (6cyl-1qt, 8cyl-1½qt)
263	Auxiliary seat (Sedan Delivery)
302	Turboglide transmission
313	Powerglide transmission
315	Overdrive transmission
320	Electric windshield wipers
324	Power steering
326	Generator (40 amp)
335	Seat cushion, foam rubber front (El Camino)

338	Generator (35 amp)	576,573	Special Super Turbo-Thrust, 315hp
345	Battery, heavy duty	576	Special Turbo-Thrust engine, 305hp
347	De Luxe equipment (El Camino)	577	Special Turbo-Thrust, 320hp
348	Steering wheel (deluxe)	577,574	Special Super Turbo-Thrust, 335hp
345	Battery, HD	578	Ramjet Fuel Injection, 250hp
378	Generator, (50 amp)	578	Special Ramjet Fuel Injection, 290hp
398	Tinted glass (E-Z-Eye)		
412	Vacuum power brake	591	Air cleaner, oil bath
417	Positive crankcase ventilation	592	Oil filter (2qt)
427	Padded instrument panel	593	HD rear coil springs
482	Full width seat (Sedan Delivery)	676	Positraction rear axle
576	Turbo-Thrust engine, 250hp	685	Transmission, 4-speed manual (El Camino)
576,573	Super Turbo-Thrust engine, 280hp		

1959 Apache 31 Suburban Carryall.

Facts

Changes were few on the Massive-Functional—styled 1959 trucks. The hood emblem was different, and the side emblems were now located above the front fenders at the fender crease. New two-tone coloring gave the trucks a different appearance—this was the Color-break two-toning. The upper body and the side spears of the pickup box were painted the main color, and the body and the rear cab pillars were painted Bombay Ivory. The interior also got new colors as well.

In the engine compartment, besides there was a Maximum Economy option was available on the 235ci six on 3100 and 3200 Series trucks. This version was rated at 110hp and was coupled to an axle that contained a 3.38:1 final drive ratio. Fuel benefits were achieved by a lower-lift camshaft and smaller carburetor. Also new was the Positraction rear axle.

The Custom Cab equipment included a foam bench seat, chrome dash knobs and window moldings, dual sun visors, dual armrests, and a cigar lighter.

The sedan delivery was still available but was also joined by the new El Camino pickup, which was a car-truck hybrid. Sharing the Chevrolet passenger-car line chassis and body, the El Camino could be optioned out with most passenger-car options and all passenger-car and truck engines. In stock form, the El Camino came with a Biscayne-level interior and Bel Air exterior trim. In stock form, the El Camino did not come with dual sun visors, armrests, or a padded dash, but it could so be optioned.

The top engine option for the El Camino was the Special Super Turbo Thrust big-block rated at 335hp. It was equipped with a 3x2bbl triple two-barrel intake setup.

Chapter 16

1960 Trucks

Production

1170, 1270 2dr Sedan Delivery	5,266
1180, 1280 2dr Sedan Pickup	
(El Camino)	23,837
Calendar year—1960	
Six cyl.	339,339
V-8	54,785
Total	394,014

Serial numbers

Description

0C144A000001

0—Last digit of model year, 0—1960

C—Chassis type, C—conventional, K—four-wheel drive, G—Sedan & Pickup Delivery 6cyl, H—Sedan & Pickup Delivery 8cyl

14—Series, 11—Sedan & Pickup Delivery 6cyl, 12—Sedan & Pickup Delivery 8cyl, 14&15—Truck ½ ton, 25—Truck ¾ ton, 36—Truck 1-ton

4—Model type, 2—chassis-cowl, 3—chassis-cab, 4—pickup, 5—panel, 6—carryall, 9—stake

A—Assembly Plant Code: A—Atlanta GA, B—Baltimore MD, F—Flint MI, J—Janesville WI, K—Kansas City MO, L—Los Angeles CA, N—Norwood OH, O—Oakland CA, S—St. Louis MO, T—Tarrytown NY, W—Willow Run MI

000001—Consecutive Sequence Number

Location: On plate attached to left door hinge post. Chassis-cowls have plate attached to left side of dash.

Model Identification

C1004

C—Chassis type, C—conventional, K—four-wheel drive, G—Sedan & Pickup Del. 6cyl, H—Sedan & Pickup Del. 8cyl

10—Truck series, 10's—½ ton, 20's—¾ ton, 30's—1 ton

04—Body/truck type, 02—Flat face cowl, 03—Chassis-cab, 04—Stepside pickup, 05—Panel, 06—Carryall w/panel rear doors, 09—Stake, 12—Windshield-cowl, 16—Carryall w/gate, 34—Fleetside pickup, 70—Sedan Delivery, 80—Pickup Delivery (El Camino)

Model, Wheelbase & GVW

Model Number and Description	Wheelbase (in)	GVW
1100/1200 Series		
1170 Sedan Delivery, 6cyl	119	4,900
1270 Sedan Delivery, 8cyl	119	4,900
1180 El Camino 6cyl	119	4,900
1280 El Camino 8cyl	119	4,900
1000 Series		
C1402 Chassis & cowl	115	5,200
C1403 Chassis and cab	115	5,200
C1404 Stepside pickup	115	5,200
C1405 Panel delivery	115	5,200
C1406 Suburban carryall w/rear doors	115	5,200
C1412 Chassis & cowl	115	5,200
C1416 Suburban carryall w/end-gate	115	5,200
C1434 Fleetside pickup	115	5,200
C1503 Chassis & cab	127	5,200
C1504 Stepside pickup	127	5,200
C1434 Fleetside pickup	127	5,200
K1403 Chassis and cab, 4WD	115	5,600
K1404 Stepside pickup 4WD	115	5,600
K1405 Panel delivery, 4WD	115	5,600
K1434 Fleetside pickup 4WD	115	5,600
K1406 Suburban carryall w/rear gate 4WD	115	5,600
K1416 Suburban carryall w/rear doors 4WD	115	5,600
2000 Series		
C2502 Chassis & cowl	127	7,500
C2503 Chassis & cab	127	7,500
C2504 Stepside pickup	127	7,500
C2509 Stake bed	127	7,500
C2534 Fleetside pickup	127	7,500
C2512 Chassis & cowl (windshield)	127	7,500
K2503 Chassis & cab 4WD	127	7,600
K2504 Stepside pickup 4WD	127	7,600
C2534 Fleetside pickup 4WD	127	7,600
3000 Series		
C3602 Chassis & cowl	133	7,800
C3603 Chassis & cab	133	10,000
C3604 Stepside pickup	133	10,000
C3605 Panel	133	7,800
C3609 Stake bed	133	10,000
C3612 Chassis & cowl (windshield)	133	10,000

Engine & Transmission Suffix Codes

1100/1200 Series

A, AE—235ci I-6 1bbl 135hp—manual trans
AF, AG—235ci I-6 1bbl 135hp—manual trans
AJ, AK—235ci I-6 1bbl 135hp—manual trans
AM, AZ—235ci I-6 1bbl 135hp—manual trans

B, BE—235ci I-6 1bbl 135hp—Powerglide
BG, BH—235ci I-6 1bbl 135hp—Powerglide
C, CF—283ci V-8 2bbl 170hp—manual trans
CD, CG—283ci V-8 2bbl 170hp—o/d manual trans
CL—283ci V-8 4bbl 270hp—manual trans
CM—283ci V-8 4bbl 230hp—manual trans
D, DK—283ci V-8 2bbl 170hp—Powerglide
DB, DM—283ci V-8 4bbl 230hp—Powerglide
E, EG—283ci V-8 2bbl 170hp—Turboglide
EB, EJ—283ci V-8 4bbl 230hp—Turboglide
F—348ci V-8 4bbl 250hp—manual trans
FA—348ci V-8 3x2bbl 280hp—manual trans
FE—348ci V-8 3x2bbl 320hp—manual trans
FG—348ci V-8 4bbl 320hp—manual trans
FH—348ci V-8 4bbl 335hp—manual trans
FJ—348ci V-8 4bbl 335hp—manual trans
G—348ci V-8 4bbl 250hp—Powerglide
GB—348ci V-8 3x2bbl 280hp—Powerglide
GD—348ci V-8 4bbl 280hp—Powerglide
H—348ci V-8 4bbl 250hp—Turboglide
HA—348ci V-8 3x2bbl 280hp—Turboglide

C-10, C-20 Series
J—235ci I-6 1bbl 135hp—manual trans
JB—235ci I-6 1bbl 135hp—Powerglide
JC—235ci I-6 1bbl 135hp—manual trans
M—283ci V-8 2bbl 160hp—manual trans
MA—283ci V-8 2bbl 160hp—Powerglide

C-30 Series
J—235ci I-6 1bbl 135hp—manual trans
M—283ci V-8 2bbl 160hp—Powerglide

K-10, K-20 Series
JC—235ci I-6 1bbl 135hp—manual trans
M—283ci V-8 2bbl 160hp—manual trans

Transmission Codes

Code	Type	Plant
B	Turboglide	Toledo
C	Powerglide	Cleveland
H	3 speed HD	Saginaw
S	3 speed O/D	Saginaw
W	4 speed	Warner Gear

Axle Identification

Code	Ratio

1100/1200 Series

AW, BW, AX, BX	3.08:1
AB, BB, AM, BM	3.36:1
AA, BA, AK, BK	3.55:1
FE, BE, FH, BH	3.70:1
AC, BC, AL, BL	4.11:1

1000 Series

MB	3.38:1
MA, MC, MD, ME, MF, MG	3.90:1

2000 Series

MH, MJ, MK	4.57:1

3000 Series

PA, PB	5.14:1

Exterior Color Codes

1100/1200 Series

Tuxedo Black	900
Cascade Green	903
Jade Green	905
Horizon Blue	910
Royal Blue	912

Tasco Turquoise	915*
Suntan Copper	920*
Roman Red	923
Crocus Cream	925*
Ermine White	936
Fawn Beige	938*
Sateen Sliver	940*
Shadow Gray	941
*El Camino	

1000/2000/3000 Series

Jet Black	700
Neptune Green	703
Hemlock Green	705
Brigade Blue	707
Marlin Blue	708
Tartan Turquoise	710
Klondike Gold	713
Cardinal Red	714
Grenadier Red	715
Omaha Orange	716
Golden Yellow	718
Yukon Yellow	719
Pure White	721
Garrison Gray	723

Regular Production Options
1100/1200 Series

1170 Sedan Delivery, 6cyl	$2,361
1180 El Camino 6cyl	2,366
1280 El Camino 8cyl	2,473

1000 Series

C1402 Chassis & cowl	1,623
C1403 Chassis and cab	1,877
C1404 Stepside pickup	1,991
C1405 Panel delivery	2,308
C1406 Suburban carryall w/rear doors	2,690
C1416 Suburban carryall w/end-gate	2,723
C1434 Fleetside pickup	2,007
C1503 Chassis & cab	1,914
C1504 Stepside pickup	2,028
C1434 Fleetside pickup	2,044
K1403 Chassis and cab, 4WD	2,640
K1404 Stepside pickup 4WD	2,754
K1405 Panel delivery, 4WD	3,071
K1434 Fleetside pickup 4WD	2,770
K1406 Suburban carryall w/rear doors 4WD	3,453
K1416 Suburban carryall w/rear gate 4WD	3,486

2000 Series

C2502 Chassis & cowl	1,795
C2503 Chassis & cab	2,069
C2504 Stepside pickup	2,173
C2509 Stake bed	2,264
C2534 Fleetside pickup	2,189
K2503 Chassis & cab 4WD	2,849
K2504 Stepside pickup 4WD	2,693
C2534 Fleetside pickup 4WD	2,979

3000 Series

C3602 Chassis & cowl	1,952
C3603 Chassis & cab	2,216
C3604 Stepside pickup	2,350
C3605 Panel	2,775
C3609 Stake bed	2,463

1000/2000/3000 Series

Option Number

103	Radio, manual
112	Heater
115	Heater, recirculating
200	Shock absorbers, HD
210	Exterior mirror
211	Rear shock absorber shields
212	Power brakes
218	Rear bumper
227	Heavy-duty clutch
230	Platform body
237	Oil filter (1 qt.)
241	Governor
253	Heavy duty front springs
254	Heavy duty rear springs (8 leaf)
256	Heavy duty radiator
258	Foam rubber seat cushion
263	Auxiliary seat
267	Heavy duty rear springs (13 leaf)
313	Powerglide automatic transmission
314	Hydra-matic 4-speed automatic transmission
315	Overdrive transmission
316	HD 3-speed transmission
318	4-speed transmission
320	Electric windshield wipers
324	Power steering
336	Generator, 40 amp
338	Generator, 35 amp
341	Side mounted spare wheel carrier
345	Heavy duty battery
350	Power steering (6cyl)
351	Power steering (V-8)
378	Generator, 50 amp

379	Generator, 50 amp, HD
393	Chrome equipment
394	Full view rear window
395	Left hand door and side mounted spare wheel key lock
396	Tire lock
398	E-Z Eye glass
408	Trademaster V-8 engine
423	Running boards
431	Custom panel
587	Generator, heavy duty
591	Air cleaner, oil bath
592	Oil filter (2qt)
599	Right door & tire lock
675	Positraction rear axle
683	Free wheeling hubs
695	Bostrom seat

1100/1200 Sedan Delivery/El Camino

103	Radio, manual control
104	Radio, push-button control
109	Windshield washers
110	Air conditioning
112	Heater
117	Wheel discs
129	Mirror, inside, non-glare
216	Air cleaner, oil bath
220	Dual exhausts
227	Heavy-duty clutch
237	Oil filter (6cyl-1qt, 8cyl-1$\frac{1}{2}$qt)
263	Auxiliary seat (Sedan Delivery)

302	Turboglide transmission
313	Powerglide transmission
315	Overdrive transmission
324	Power steering
326	Generator (40 amp)
333	Electric windshield wipers
335	Seat cushion, foam rubber front
338	Generator (35 amp)
345	Battery, heavy duty
347	De Luxe equipment
348	Steering wheel (deluxe)
345	Battery, HD
378	Generator, (50 amp)
398	Tinted glass (E-Z-Eye)
410	Super Turbo-Fire engine, 230hp
412	Brakes, vacuum power
427	Padded instrument panel
482	Full width seat (Sedan Delivery)
576	Turbo-Thrust engine, 250hp
576/573	Super Turbo-Thrust engine, 280hp
576/573	Special Super Turbo-Thrust, 315hp
576	Special Turbo-Thrust engine, 305hp
577	Special Turbo-Thrust, 320hp
577/574	Special Super Turbo-Thrust, 335hp
581	Carburetor, economy
593	HD rear coil springs
675	Positraction rear axle
685	Transmission, 4-speed manual (El Camino)

Facts

The year 1960 began a new era for Chevrolet trucks, known as the V-8 era. This period extended to 1966. The 1960 trucks were redesigned, acquiring a squarer appearance. Although they were 7in lower in height, they were larger in every other dimension. The most noticeable styling effect was two prominent hood pods, resembling air intakes. In the center of the hood was the familiar Chevrolet bow-tie emblem. If the truck was equipped with a V-8 engine, the bow tie was backed by a large V ornament.

Several mechanical changes were made. The frame, though similar to the previous design, was buttressed by an additional center X member cross-member, giving the chassis extra rigidity. The front suspension was totally new. It was an independent setup that used ball joints and torsion bars instead of the more common leaf springs. At the rear, 1/2- and 3/4-ton models rode on leaf springs; 1-ton models still used leaf springs. The use of coil springs and an independent front suspension resulted in a better ride and easier steering on trucks that were not equipped with power steering.

The powertrain choice was basically a carry-over from 1959: the 235ci Thriftmaster inline six rated at 135hp and the 283ci Trademaster V-8 at 160hp. Also available on the C-10 was the Thriftmaster Economy option, in which the 235ci six was rated at 110hp. The optional four-speed Hydra-matic automatic was not available on the 1/2- and 3/4-ton models, being replaced by the two-speed Powerglide automatic.

Four-wheel-drive models came standard with a three-speed manual, and the four-speed manual transmission was optional. Four-wheel-drive trucks came with a Spicer front axle and used leaf springs on each wheel. The center X frame additional center cross-member was eliminated on the four-wheel-drive models.

The light trucks were no longer called Apaches. The Apache was now a trim package that consisted of chrome grille bars and front bumper, and special fender emblems and bodyside moldings. Apache-equipped trucks also got the area behind the rear cab pillar and window painted a contrasting color.

In standard form, the trucks were equipped with the Ivory White painted front bumper, grille, and hood vents painted Ivory White. A chrome-plated bumper and grille were optional. The standard trim level was called Deluxe, and a Custom trim treatment was optional. The Custom Cab Package included a chrome front grille and bumper; chrome vent and door window moldings; bright windshield, upper—rear quarter, and beltline moldings; with and black Custom script. The Custom also included a passenger's-side visor, a left-side armrest, a left-side door lock, simulated door trim, a cigar lighter, chrome dash knobs, a chrome Chevrolet emblem on the dash, and better padded cloth seats faced with charcoal vinyl. faced seats.

The Suburban Carryall continued in the same basic configuration as before. It was available either with panel-type rear doors or with a tailgate. The side windows were of the sliding variety. Three seats were installed: the driver's bench with a folding passenger section, a two-passenger center seat, and a three-passenger rear seat. The center and rear seats could be interchanged. The panel truck came with a single driver's seat and with the floor covered in wood. Metal sliding strips were attached to the floor.

Two-tone paints were available on the Suburban and anel truck. The front roof section, cab pillars, hood, and body midriff sections were painted Ivory White, snf the rear roof section, side window pillars, rear window area, and lower body were painted body color.

In all, 165 different truck models were available.

The El Camino took on the passenger-car styling features. This was, however, the last year for the El Camino to be offered on the full-size passenger-car platform. The El Camino did not return in 1961, but was reintroduced in 1964 on the Chevelle intermediate line.

This was the last year for the sedan delivery.

Chapter 17

1961 Trucks

Production

Corvair R1244 Loadside	2,475
Corvair R1254 Rampside	10,787
Calendar year—1961	
Six cyl.	294,813
V-8	48,391
Diesel	84
Total	342,658

Serial numbers
Description
1C144A000001

1—Last digit of model year, 1—1961

C—Chassis type, C—conventional, K—four-wheel drive, R—Corvair 95

14—Series, 12—Corvair 95, 14&15—Truck ½ ton, 25—Truck ¾ ton, 36—Truck 1-ton

4—Model type, 2—Chassis-cowl, 3—Chassis-cab, 4—Pickup, 5—Panel, 6—Carryall, 9—Platform stake

A—Assembly Plant Code: A—Atlanta GA, B—Baltimore MD, F—Flint MI, J—Janesville WI, K—Kansas City MO, L—Los Angeles CA, N—Norwood OH, O—Oakland CA, S—St. Louis MO, T—Tarrytown NY, W—Willow Run MI

000001—Consecutive Sequence Number

Location: On plate attached to left door hinge post. Chassis-cowls have plate attached to left side of dash.

Model Identification
C1004

C—Chassis type, C—conventional, K—four-wheel drive, R—Corvair 95

10—Truck series, 10's—½ ton, 20's—¾ ton, 30's—1 ton

04—Body/truck type, 02—Flat face cowl, 03—Chassis-cab, 04—Stepside pickup, 05—Panel, 06—Carryall w/panel rear doors, 09—Stake, 12—Windshield-cowl, 16—Carryall w/gate, 34—Fleetside pickup, 44—Loadside pickup, 54—Rampside pickup

Model, Wheelbase & GVW

Model Number and Description	Wheelbase (in)	GVW
Corvair 95		
R1244 Loadside pickup	95	4,600
R1254 Rampside pickup	95	4,600

1000 Series

C1402	Chassis and cowl	115	4,600
C1403	Chassis and cab	115	4,600
C1404	Stepside pickup	115	4,600
C1405	Panel delivery	115	4,600
C1406	Suburban carryall w/rear doors	115	4,600
C1412	Chassis & cowl	115	4,600
C1416	Suburban carryall w/end-gate	115	4,600
C1434	Fleetside pickup	115	5,300
C1503	Chassis & cab	127	5,600
C1504	Stepside pickup	127	4,600
C1534	Fleetside pickup	127	4,600
C1434	Fleetside pickup	127	5,300
K1403	Chassis and cab, 4WD	115	5,300
K1404	Stepside pickup 4WD	115	5,600
K1405	Panel delivery, 4WD	115	5,300
K1434	Fleetside pickup 4WD	115	5,300
K1406	Suburban carryall w/rear gate 4WD	115	5,300
K1416	Suburban carryall w/rear doors 4WD	115	5,300

2000 Series

C2502	Chassis & Cowl	127	7,500
C2503	Chassis & cab	127	7,500
C2504	Stepside pickup	127	7,500
C2509	Stake bed	127	7,500
C2534	Fleetside pickup	127	7,500
C2512	Chassis & cowl (windshield)	127	7,500
K2503	Chassis & cab 4WD	127	7,600
K2504	Stepside pickup 4WD	127	7,600
C2534	Fleetside pickup 4WD	127	7,600

3000 Series

C3602	Chassis & cowl	133	7,800
C3603	Chassis & cab	133	7,800
C3604	Stepside pickup	133	7,800
C3605	Panel	133	7,800
C3609	Stake bed	133	7,800
C3612	Chassis & cowl (windshield)	133	7,800

Engine & Transmission Suffix Codes

Corvair 95

V—145ci F-6 2-1bbl 80hp—manual trans
W—145ci F-6 2-1bbl 80hp—Powerglide

C-10, C-20, C-30 Series

J—235ci I-6 1bbl 135hp—manual trans
JB—235ci I-6 1bbl 135hp—Powerglide
JC—235ci I-6 1bbl 135hp—manual trans
M—283ci V-8 2bbl 160hp—manual trans
MA—283ci V-8 2bbl 160hp—Powerglide

K-10, K-20 Series

JC—235ci I-6 1bbl 135hp—manual trans
M—283ci V-8 2bbl 160hp—manual trans

Transmission Codes

Code	Type	Plant
B	Turboglide	Toledo
C	Powerglide	Cleveland
H	3 speed HD	Saginaw
S	3 speed O/D	Saginaw
W	4 speed	Warner Gear

Axle Identification

Code	Ratio

Corvair
BL, BY	3.89:1

C-10/K-20 Series
MB	3.38:1
MA, MC, MD, ME, MF, MG	3.90:1

C-20/K-20 Series
MH, MJ, MK	4.57:1

C-30 Series
PA, PB	5.14:1

Exterior Color Codes

C-10/C-20/C-30 Series
Jet Black	700
Neptune Green	703
Woodland Green	705
Brigade Blue	707
Balboa Blue	708
Grenadier Red	715
Flaxen Yellow	718
Yukon Yellow	719
Cameo White	726
Pure White	721
Woodsmoke Blue	723
Romany Maroon	724
Tahiti Coral	725
Tampico Turquoise	727
Cardinal Red	740
Omaha Orange	742

Regular Production Options

Corvair Series
R1244	Loadside pickup	$2,135
R1254	Rampside pickup	2,080

1000 Series

C1403	Chassis and cab	1,875
C1404	Stepside pickup	1,990
C1405	Panel delivery	2,310
C1406	Suburban carryall w/rear doors	2,670
C1416	Suburban carryall w/end-gate	2,704
C1434	Fleetside pickup	2,005
C1503	Chassis & cab	1,915
C1504	Stepside pickup	2,030
C1534	Fleetside pickup	2,045
C1434	Fleetside pickup	2,005
K1403	Chassis and cab, 4WD	2,555
K1404	Stepside pickup 4WD	2,670
K1405	Panel delivery, 4WD	2,986
K1434	Fleetside pickup 4WD	2,685
K1406	Suburban carryall w/rear doors 4WD	3,345
K1416	Suburban carryall w/rear gate 4WD	3,379

2000 Series

C2503	Chassis & cab	2,060
C2504	Stepside pickup	2,175
C2509	Stake bed	2,265
C2534	Fleetside pickup	2,190
K2503	Chassis & cab 4WD	2,765
K2504	Stepside pickup 4WD	2,875
C2534	Fleetside pickup 4WD	2,875

3000 Series

C3603	Chassis & cab	2,216
C3604	Stepside pickup	2,350
C3605	Panel	2,775
C3609	Stake bed	2,463

1000/2000/3000 Series

Option Number

105	Directional signals
115	Heater, recirculating
123	Radio, manual
210	Exterior mirror
211	Rear shock absorber shields
212	Power brakes
213	Heavy duty shock absorbers
218	Painted rear bumper
223	Heavy-duty clutch
230	Platform body
237	Oil filter (1 qt.)
241	Governor
253	Heavy duty front springs
254	Heavy duty rear springs (8 leaf)
256	Heavy duty radiator
258	Foam rubber seat cushion
263	Auxiliary seat

1961 Corvair pickup. Paul G. McLaughlin

1961 Chevrolet Apache 1/2 ton pickup in photographer's studio. Don Bunn

Facts

Few changes were made on the 1961 models. The Chevrolet lettering was relocated to the center of the grille, and trim wings were placed on the sides of each parking light. The hubcaps had a flat center section, which was stamped with the bow-tie emblem. The side emblems were relocated to above the body-side crease line.

In the interior, the brake and clutch pedals were suspended beneath the dash, rather than protruding from the floor. The interior also got a different upholstery pattern.

The Thriftmaster 110hp six-cylinder engine was discontinued, but otherwise, the drivetrain was unchanged.

Two Corvair-based pickup trucks were introduced, in Loadside and Rampside forms. Both were powered with a rear-mounted 80hp air-cooled flat-four-cylinder engine. All four wheels were independently sprung.

The Loadside version came with panel doors behind the cab. The Rampside used a hinged door on the passenger's side of the cab, which dropped to the ground to become a loading ramp. These pickups changed little during the time they were available.

Chapter 18

1962 Trucks

Production
Corvair R1244 Loadside	369
Corvair R1254 Rampside	4,102
Calendar year	Total
1962	396,819

Serial numbers

Description
2C144A000001*

2—Last digit of model year, 2—1962

C—Chassis type, C—conventional, K—four-wheel drive, R—Corvair 95

14—Series, 12—Corvair 95, 14&15—Truck ½ ton, 25—Truck ¾ ton, 36—Truck 1-ton

4—Model type, 2—Chassis-cowl, 3—Chassis-cab, 4—Pickup, 5—Panel, 6—Carryall, 9—Platform stake

A—Assembly Plant Code: A—Atlanta GA, B—Baltimore MD, F—Flint MI, J—Janesville WI, K—Kansas City KS, L—Los Angeles CA, N—Norwood OH, O—Oakland CA, S—St. Louis MO, T—Tarrytown NY, W—Willow Run MI

000001—Consecutive Sequence Number

* A prefix "S" indicates a C-30 Series with 7,800 GVW rating

Location: On plate attached to left door hinge post. Chassis-cowls have plate attached to left side of dash.

Model Identification
C1004

C—Chassis type, C—conventional, K—four-wheel drive, R—Corvair 95

10—Truck series, 10's—½ ton, 20's—¾ ton, 30's—1 ton

04—Body/truck type, 02—Flat face cowl, 03—Chassis-cab, 04—Stepside pickup, 05—Panel, 06—Carryall w/panel rear doors, 09—Stake, 12—Windshield-cowl, 16—Carryall w/gate, 34—Fleetside pickup, 44—Loadside pickup, 54—Rampside pickup

Model, Wheelbase & GVW

Model Number and Description	Wheelbase (in)	GVW
Corvair 95		
R1244 Loadside pickup	95	4,600
R1254 Rampside pickup	95	4,600

90

1000 Series

C1402	Chassis and cowl	115	4,600
C1403	Chassis and cab	115	4,600
C1404	Stepside pickup	115	4,600
C1405	Panel delivery	115	4,600
C1406	Suburban carryall w/rear doors	115	4,600
C1412	Chassis & cowl	115	4,600
C1416	Suburban carryall w/end-gate	115	4,600
C1434	Fleetside pickup	115	5,300
C1503	Chassis & cab	127	5,600
C1504	Stepside pickup	127	4,600
C1534	Fleetside pickup	127	4,600
C1434	Fleetside pickup	127	5,300
K1403	Chassis and cab, 4WD	115	5,300
K1404	Stepside pickup 4WD	115	5,600
K1405	Panel delivery, 4WD	115	5,300
K1434	Fleetside pickup 4WD	115	5,300
K1406	Suburban carryall w/rear gate 4WD	115	5,300
K1416	Suburban carryall w/rear doors 4WD	115	5,300

2000 Series

C2502	Chassis & cowl	127	7,500
C2503	Chassis & cab	127	7,500
C2504	Stepside pickup	127	7,500
C2509	Stake bed	127	7,500
C2534	Fleetside pickup	127	7,500
C2512	Chassis & cowl (windshield)	127	7,500
K2503	Chassis & cab 4WD	127	7,600
K2504	Stepside pickup 4WD	127	7,600
K2534	Fleetside pickup 4WD	127	7,600

3000 Series

C3602	Chassis & cowl	133	7,800
C3603	Chassis & cab	133	7,800
C3604	Stepside pickup	133	7,800
C3605	Panel	133	7,800
C3609	Stake bed	133	7,800
C3612	Chassis & cowl (windshield)	133	7,800

Engine & Transmission Suffix Codes

Corvair 95

V—145ci F-6 2-1bbl 80hp—manual trans
W—145ci F-6 2-1bbl 80hp—Powerglide

C-10, C-20 Series

J—235ci I-6 1bbl 135hp—manual trans
JB—235ci I-6 1bbl 135hp—Powerglide
JC—235ci I-6 1bbl 135hp—manual trans
—261ci I-6 1bbl 150hp—manual trans
M—283ci V-8 2bbl 160hp—manual trans
MA—283ci V-8 2bbl 160hp—Powerglide

C-30 Series

J—235ci I-6 1bbl 135hp—manual trans
JC—235ci I-6 1bbl 135hp—manual trans
—261ci I-6 1bbl 150hp—manual trans
M—283ci V-8 2bbl 160hp—manual trans

K-10, K-20 Series

JC—235ci I-6 1bbl 135hp—manual trans
—261ci I-6 1bbl 150hp—manual trans
M—283ci V-8 2bbl 160hp—manual trans

Transmission Codes

Code	Type	Plant
B	Powerglide	Toledo (Corvair)
C	Powerglide	Cleveland
M	3-4 speed	Muncie
S	3 speed	Saginaw
W	3 speed HD	Warner Gear

Axle Identification

Code	Ratio

Corvair

HE, HF, HL, HM	3.89:1

C-10, K-20 Series

MB, ME	3.38:1
MA, MC, MD, MF, MG	3.90:1
MT, MU	4.11:1

C-20, K-20 Series

MH, MJ, MP, MQ	4.57:1

C-30 Series

PA, PB PL, PN	5.14:1

Exterior Color Codes

C-10/C-20/C-30 Series

Jet Black	500
Seamist Green	502
Glenwood Green	503
Woodland Green	505
Brigade Blue	507
Balboa Blue	508
Crystal Turquoise	510
Cardinal Red	514
Omaha Orange	516
Yuma Yellow	519
Pure White	521
Georgian Gray	522
Cameo White	526
Desert Beige	528

Regular Production Options

Corvair Series

R1244	Loadside pickup	$2,135
R1254	Rampside pickup	2,080

1000 Series

C1403	Chassis and cab	1,875
C1404	Stepside pickup	1,990
C1405	Panel delivery	2,310
C1406	Suburban carryall w/rear doors	2,670
C1416	Suburban carryall w/end-gate	2,704
C1434	Fleetside pickup	2,005
C1503	Chassis & cab	1,915
C1504	Stepside pickup	2,030
C1534	Fleetside pickup	2,045
C1434	Fleetside pickup	2,005
K1403	Chassis and cab, 4WD	2,555
K1404	Stepside pickup 4WD	2,670
K1405	Panel delivery, 4WD	2,986
K1434	Fleetside pickup 4WD	2,685
K1406	Suburban carryall w/rear doors 4WD	3,345
K1416	Suburban carryall w/rear gate 4WD	3,379

2000 Series

C2503	Chassis & cab	2,060
C2504	Stepside pickup	2,175
C2509	Stake bed	2,265
C2534	Fleetside pickup	2,190
K2503	Chassis & cab 4WD	2,765
K2504	Stepside pickup 4WD	2,875
C2534	Fleetside pickup 4WD	2,875

3000 Series

C3603	Chassis & cab	2,216
C3604	Stepside pickup	2,350
C3605	Panel	2,775
C3609	Stake bed	2,463

1000/2000/3000 Series

Option Number

112	Deluxe heater
115	Heater, recirculating
123	Radio, manual
124	Radiator fan
205	Single speed rear axle 4.11:1
210	Exterior rearview mirror
212	Power brakes
213	Heavy duty shock absorbers
215	Single speed rear axle 3.38:1
218	Painted rear bumper

223 Heavy-duty clutch
241 Governor
243 Special crankcase ventilation
254 Heavy duty rear springs (8 leaf)
256 Heavy duty radiator
258 Full depth foam seat
264 Auxiliary seat
266 Tachometer
267 Auxiliary rear springs
269 Third seat—Carryall
293 261ci six cylinder engine
311 Powerglide automatic
      transmission
316 HD 3-speed transmission
318 4-speed transmission
320 42 amp Delcotron alternator
338 Generator, 35 amp
341 Side mounted spare wheel
      carrier
355 Two speed windshield wipers
356 Heavy duty battery

370 Laminated glass
371 Maximum economy equipment
379 7800 lb. GVW plate
391 Mechanical jack
393 Custom chrome equipment
394 Full view rear window
395 Right door lock
408 Trademaster V-8 engine
411 Soft ray glass
432 Custom appearance option
433 Custom comfort option
443 52 amp Delcotron alternator
448 62 amp Delcotron alternator
472 Extended range fuel tank
591 Oil bath air cleaner
592 Oil filter (2qt)
677 No-Spin rear axle
680 Positraction rear axle
683 Free wheeling front hubs
695 Bostrom seat

Facts

The 1962 trucks were facelifted. The twin hood pods were eliminated, giving the front end a cleaner look, and the dual headlights were replaced by a single 7in unit. The Chevrolet lettering was relocated to the bottom bar of the grille.

An additional optional engine was available on the 1/2- and 3/4-ton models. This was a larger version of the 235ci inline six, displacing 261ci and rated at 150hp. The other engines were a 135hp six and a 160hp 283ci V-8.

Four optional trim packages were available. The Custom comfort option included 6in foam-padded seats, distinctive cloth-and-vinyl trim, a left-door armrest, passenger's-side sun visor and door lock, and a cigar lighter. The Custom chrome option included front chrome bumper and hubcaps. The Custom appearance option included a bright grille, bright windshield molding, and upper rear quarter panel; Custom trim plates; and in the interior, a steering wheel with a chrome horn ring, brightmetal trim on the dash, and two-tone door panels. The Custom side molding option got a full-length trim spear. It was available only on the Fleetside models.

As in previous years, the trucks did not come with a rear bumper. This was optional and could be had either painted or chrome-plated.

The Corvair pickups continued almost unchanged but could now be equipped with a Positraction axle.

Chapter 19

1963 Trucks

Production

Corvair R1254 Rampside	2,046
Calendar year	Total
1963	483,119

Serial numbers

Description

3C144A000001*

3—Last digit of model year, 3—1963

C—Chassis type, C—conventional, K—four-wheel drive, R—Corvair 95

14—Series, 12—Corvair 95, 14&15—Truck ½ ton, 25—Truck ¾ ton, 36—Truck 1-ton

4—Model type, 2—Chassis-cowl, 3—Chassis-cab, 4—Pickup, 5—Panel, 6—Carryall, 9—Platform stake

A—Assembly Plant Code: A—Atlanta GA, B—Baltimore MD, F—Flint MI, G—Framingham MA, J—Janesville WI, K—Kansas City MO, L—Los Angeles CA, N—Norwood OH, O—Oakland CA, P—Pontiac MI, S—St. Louis MO, T—Tarrytown NY, W—Willow Run MI

000001—Consecutive Sequence Number

* A prefix "S" indicates a C-30 Series with 7,800 GVW rating

Location: On plate attached to left door hinge post. Chassis-cowls have plate attached to left side of dash.

Model Identification

C1004

C—Chassis type, C—conventional, K—four-wheel drive, R—Corvair 95

10—Truck series, 10's—½ ton, 20's—¾ ton, 30's—1 ton

04—Body/truck type, 02—Flat face cowl, 03—Chassis-cab, 04—Stepside pickup, 05—Panel, 06—Carryall w/panel rear doors, 09—Stake, 12—Windshield-cowl, 16—Carryall w/gate, 34—Fleetside pickup, 44—Loadside pickup, 54—Rampside pickup

Model, Wheelbase & GVW

Model Number and Description	Wheelbase (in)	GVW
Corvair 95		
R1254 Rampside pickup	95	4,600

C-10 Series

C1402	Chassis and cowl	115	4,600
C1403	Chassis and cab	115	4,600
C1404	Stepside pickup	115	4,600
C1405	Panel delivery	115	4,600
C1406	Suburban carryall w/rear doors	115	4,600
C1412	Chassis & cowl	115	4,600
C1416	Suburban carryall w/end-gate	115	4,600
C1434	Fleetside pickup	115	5,300
C1503	Chassis & cab	127	5,600
C1504	Stepside pickup	127	4,600
C1534	Fleetside pickup	127	4,600
C1434	Fleetside pickup	127	5,300
K1403	Chassis and cab, 4WD	115	5,300
K1404	Stepside pickup 4WD	115	5,600
K1405	Panel delivery, 4WD	115	5,300
K1434	Fleetside pickup 4WD	115	5,300
K1406	Suburban carryall w/rear gate 4WD	115	5,300
K1416	Suburban carryall w/rear doors 4WD	115	5,300
K1503	Chassis and cab 4WD	127	5,600
K1504	Stepside pickup 4WD	127	5,600
K1534	Fleetside pickup 4WD	127	5,600

C-20 Series

C2502	Chassis & cowl	127	7,500
C2503	Chassis & cab	127	7,500
C2504	Stepside pickup	127	7,500
C2509	Stake bed	127	7,500
C2534	Fleetside pickup	127	7,500
C2512	Chassis & cowl (windshield)	127	7,500
K2503	Chassis & cab 4WD	127	7,600
K2504	Stepside pickup 4WD	127	7,600
K2534	Fleetside pickup 4WD	127	7,600

C-30 Series

C3602	Chassis & cowl	133	7,800
C3603	Chassis & cab	133	7,800
C3604	Stepside pickup	133	7,800
C3605	Panel	133	7,800
C3609	Stake bed	133	7,800
C3612	Chassis & cowl (windshield)	133	7,800

Engine & Transmission Suffix Codes

Corvair 95

V—145ci F-6 2-1bbl 80hp—manual trans
W—145ci F-6 2-1bbl 80hp—Powerglide

C-10, C-20 Series

N—230ci I-6 1bbl 140hp—manual trans
ND—230ci I-6 1bbl 140hp—Powerglide
NE—230ci I-6 1bbl 140hp—manual trans
PG—292ci I-6 1bbl 165hp—manual trans
PK—292ci I-6 1bbl 165hp—Powerglide
M—283ci V-8 2bbl 175hp—manual trans
MA—283ci V-8 2bbl 175hp—Powerglide

C-30 Series

NE—230ci I-6 1bbl 140hp—manual trans
PG—292ci I-6 1bbl 165hp—manual trans
M—283ci V-8 2bbl 175hp—manual trans

K-10, K-20 Series

PG—292ci I-6 1bbl 165hp—manual trans
M—283ci V-8 2bbl 170hp—manual trans

Transmission Codes

Code	Type	Plant
C	Powerglide	Cleveland
M	3 speed OD	Muncie
N	4 speed	Muncie
S	3 speed	Saginaw
T	Powerglide	Toledo (Corvair)
W	3 speed	Warner Gear

Axle Identification

Code	Ratio

Corvair

HE, HF, HL, HM	3.89:1

C-10, K-20 Series

MB, ME	3.38:1
MA, MC, MD, MF, MG	3.90:1
MT, MU	4.11:1

C-20, K-20 Series

MH, MJ, MK MP, MQ	4.57:1
MM, MV MX, MZ	4.57:1 (dual wheels)

C-30 Series

PA, PB PL, PN	5.14:1

Exterior Color Codes

C-10, C-20, C-30 Series

Jet Black	500
Seamist Jade	502
Glenwood Green	503
Woodland Green	505
Brigade Blue	507
Balboa Blue	508
Crystal Turquoise	510

Exterior Color Codes

C-10, C-20, C-30 Series

Cardinal Red	514
Omaha Orange	516
Yuma Yellow	519
Pure White	521
Georgian Gray	522
Cameo White	526
Desert Beige	528

Regular Production Options

Corvair Series

R1244	Loadside pickup	$2,136

C-10 Series

C1403	Chassis and cab	1,895
C1404	Stepside pickup	2,009
C1405	Panel delivery	2,326
C1406	Suburban carryall w/rear doors	2,620
C1416	Suburban carryall w/end-gate	2,653
C1434	Fleetside pickup	2,025
C1503	Chassis & cab	1,933
C1504	Stepside pickup	2,046
C1534	Fleetside pickup	2,062
C1434	Fleetside pickup	2,025
K1403	Chassis and cab, 4WD	2,546
K1404	Stepside pickup 4WD	2,660
K1405	Panel delivery, 4WD	2,977
K1434	Fleetside pickup 4WD	2,676
K1406	Suburban carryall w/rear doors 4WD	3,271
K1416	Suburban carryall w/rear gate 4WD	3,304

C-20 Series

C2503	Chassis & cab	2,079
C2504	Stepside pickup	2,193
C2509	Stake bed	2,284
C2534	Fleetside pickup	2,209
K2503	Chassis & cab 4WD	2,757
K2504	Stepside pickup 4WD	2,869
C2534	Fleetside pickup 4WD	2,885

C-30 Series

C3603	Chassis & cab	2,236
C3604	Stepside pickup	2,371
C3605	Panel	2,795
C3609	Stake bed	2,463

C-10, C-20, C-30 Series
Option Number

A07 Body glass
A08 R.H. side body glass
AO9 Laminated glass
A10 Panoramico Cab
A11 Tinted glass
A12 Rear door glass
A13 Side door glass
A18 Swing-out rear door glass
A24 Cab corner windows
A34 Drivers seat—Bostrom
A50 Bucket seats
A52 Custom bench seat
A54 Full width front seat
A55 Level ride seat
A56 Drivers seat—Bostrom Levelair
A57 Auxiliary seat—1 passenger
A58 Auxiliary seat—2-passenger
A59 Supplementary seat
A60 Drivers seat—national cush"N" air
A61 Auxiliary seat—stationary
A62 Front seat belt (delete)
A63 Rear seat belt (delete)
A68 Rear seat center belt
A78 Center seat
A80 Center and rear seat
A85 Shoulder harness
A94 Door safety lock
A97 Spare wheel lock
A99 Instrument panel comp.
AA2 Tinted windshield glass
AM2 Heavy duty seat
AM3 Front seat center belt
AN2 Level ride driver and auxiliary seat
AN4 Heavy duty driver and auxiliary seat
AS3 Rear seat
AS5 Shoulder harness
AU2 Cargo door lock unit
B30 Floor and toe panel carpet
B55 Full foam seat cushion
B59 Padded seat back frame
B70 Instrument panel pad
B85 Body side moulding—upper
B93 Door edge guards
B98 Side trim molding
B#2 Padded hinge pillar
BX1 Body side moulding—belt

BX2 Body side moulding—wide
CO7 Auxiliary top
C20 Single speed windshield wiper
C41 Heater—economy
C42 Heater—deluxe
C48 Less heater
C80 Air conditioning
D14 Arm rest—front door
D20 Sunshade—windshield RH
D23 Sunshade—padded LH
D29 West coast mirror—Jr
D30 West coast mirror—Sr
D32 Rear view mirror
D36 Non-glare inside mirror
D48 Front cross view mirror
D89 Body paint stripe
DG8 Outside rear view mirror—RH (fixed arm)
DG9 Outside rear view mirror—RH (swinging arm)
DH3 Outside rear view mirror—LH, RH (swinging arm)
E23 Heavy duty cab lifting torsion bar
E28 Assist handles
E56 Platform and stake rack
E57 Platform
E80 Pickup box mounting
E81 Floor board
E85 Body side door
FO2 Special heavy duty frame
FO3 Heavy duty frame
F19 Special body cross sill mounting support
F25 Frame rails—full depth heat treated
F26 Frame rails—full depth
F43 Front axle
F48 Heavy duty front axle
F51 Front and rear HD shock absorbers
F52 Front shock absorbers HD
F59 Front stabilizer
F60 Heavy duty front springs
F76 Front wheel locking hubs
F82 Front springs, HD
G45 Rear suspension, HD
G49 Rear springs, HD
G50 Heavy rear springs
G60 Auxilliary rear springs
G64 Rear spring equipment
G68 Rear shock absorbers
G70 Rear suspension—leaf spring
G77 Rear suspension—Hendrickson RT365
G80 Limited slip rear axle
HO7 Single speed rear axle
NO1 Gas tank—21 gallon

NO2	Gas tank—30 gallon	U10	Voltmeter
NO7	Gas tank—dual—37 gallon	U15	Speed warning indicator
N12	Exhaust system—single	U16	Tachometer
N13	Exhaust system—dual	U30	Oil pressure gauge
N18	Gas tank—single—37 gallon	U31	Ammeter
N20	Gas tank—single—50 gallon	U42	Directional signals
N21	Steering wheel—21″	U43	Double faced front direction signals
N60	Corrosion resistant muffler		
NE6	Step fuel tank—dual	U60	Radio—manual
NG1	Fuel tank—24½ gallon	U63	Radio—push button
PO1	Wheel trim cover	V35	Front bumper
PO3	Chrome hub cap	V37	Chrome bumper
P10	Spare wheel carrier	V38	Rear bumper—painted
P13	Wheel carrier—side mounted	V43	Rear step bumper
T60	Heavy duty battery	V46	Front bumper, chrome
T88	Parking lamps	V60	Foot tire pump
U01	Roof marker and identification lamps	V62	Hydraulic jack
		V74	Hazard warning switch
U06	Air horn	V76	Front tow hook
U08	Dual horns	VF1	Rear bumper, chrome

1963 Corvair pickup.

Facts

Stylistically, changes were few on the 1963 trucks. The headlights were located in round housings—versus oval housings in 1962—and the grille was new. The side emblems were relocated to the fenders behind the wheelwells.

Under the hood and on the chassis, however, more extensive changes were made. The old Stovebolt six, which had been around since 1929, was replaced by a modern inline six-cylinder engine. With seven main bearings, the new six used many of the reciprocating and valvetrain parts from the highly successful small-block V-8. It was available in two displacements: 230ci and 292ci, rated 140hp and 165hp respectively. Also available was an economy version of the 230ci six, rated at 125hp. The 283ci Trademaster V-8 contined, but with 175hp.

Delcotron alternators replaced generators in all applications.

The front suspension was redesigned, with coil springs replacing the previous torsion bars. This also made it possible to eliminate the central X member cross-member, frame, which had added weight. In the rear, progressive two-stage coil springs were used and the rear axles were redesigned.

Chapter 20

1964 Trucks

Production

Corvair R1254 Rampside	851
5380 El Camino 6cyl	4,430
5480 El Camino 8cyl	4,500
5580 El Camino Custom 6cyl	3,042
5680 El Camino Custom 8cyl	20,576
Total El Camino	32,548
Calendar year	Total
1964	523,791

Serial numbers

Description

4C144A000001*

4—Last digit of model year, 4—1964

C—Chassis type, C—conventional, K—four-wheel drive, R—Corvair 95

14—Series, 12—Corvair 95, 14&15—Truck ½ ton, 25—Truck ¾ ton, 36—Truck 1-ton

4—Body type, 2—Chassis-cowl, 3—Chassis-cab, 4—Pickup, 5—Panel, 6—Carryall, 9—Platform stake

A—Assembly Plant Code: A—Atlanta GA, B—Baltimore MD, F—Flint MI, G—Framingham MA, J—Janesville WI, K—Kansas City MO, L—Los Angeles CA, N—Norwood OH, O—Oakland CA, P—Pontiac MI, S—St. Louis MO, T—Tarrytown NY, W—Willow Run MI

000001—Consecutive Sequence Number

* A prefix "S" indicates a C-30 Series with 7,800 GVW rating

Location: On plate attached to left door hinge post. Chassis-cowls have plate attached to left side of dash.

El Camino Serial Number

45680A100001

4—Last digit of model year, 4—1964

36—Series, 53—El Camino 6cyl, 54—El Camino 8cyl, 55—Custom El Camino 6cyl, 56—Custom El Camino 8cyl

80—Body type, 2 door sedan pickup

A—Assembly Plant Code: A—Atlanta GA, B—Baltimore MD, H—Fremont CA, K—Kansas City MO, L—Los Angeles CA

100001—Consecutive Sequence Number

Location: On plate attached to left door hinge post.

Model Identification

C1004*

C—Chassis type, C—conventional, K—four-wheel drive, R—Corvair 95

10—Truck series, 10's—¹/₂ ton, 20's—³/₄ ton, 30's—1 ton

04—Body/truck type, 02—Flat face cowl, 03—Chassis-cab, 04—Stepside
   pickup, 05—Panel, 06—Carryall w/panel rear doors, 09—Stake,
   12—Windshield-cowl, 16—Carryall w/gate, 34—Fleetside pickup,
   44—Loadside pickup, 54—Rampside pickup

* an "S" suffix indicates a ³/₄ ton special with a 1 ton code and rating

Model, Wheelbase & GVW

Model Number and Description	Wheelbase (in)	GVW
El Camino		
5380 El Camino 6cyl	115	—
5480 El Camino 8cyl	115	—
5580 El Camino Custom 6cyl	115	—
5680 El Camino Custom 8cyl	115	—
Corvair 95		
R1254 Rampside pickup	95	4,600
C-10 Series		
R1402 Chassis and cowl	115	4,600
C1403 Chassis and cab	115	4,600
C1404 Stepside pickup	115	4,600
C1405 Panel delivery	115	4,600
C1406 Suburban carryall w/rear doors	115	4,600
C1412 Chassis & cowl	115	4,600
C1416 Suburban carryall w/end-gate	115	4,600
C1434 Fleetside pickup	115	5,300
C1503 Chassis & cab	127	5,600
C1504 Stepside pickup	127	4,600
C1534 Fleetside pickup	127	4,600
C1434 Fleetside pickup	127	5,300
K1403 Chassis and cab, 4WD	115	5,300
K1404 Stepside pickup 4WD	115	5,600
K1405 Panel delivery, 4WD	115	5,300
K1434 Fleetside pickup 4WD	115	5,300
K1406 Suburban carryall w/rear gate 4WD	115	5,300
K1416 Suburban carryall w/rear doors 4WD	115	5,300
K1503 Chassis and cab 4WD	127	5,600
K1504 Stepside pickup 4WD	127	5,600
K1534 Fleetside pickup 4WD	127	5,600
C-20 Series		
C2502 Chassis & cowl	127	6,000
C2503 Chassis & cab	127	6,000
C2504 Stepside pickup	127	6,000
C2509 Stake bed	127	6,000
C2534 Fleetside pickup	127	6,800
C2512 Chassis & cowl (windshield)	127	6,000
K2503 Chassis & cab 4WD	127	6,000
K2504 Stepside pickup 4WD	127	6,800
K2534 Fleetside pickup 4WD	127	6,800

C-30 Series

C3602	Chassis & cowl	133	7,800
C3603	Chassis & cab	133	7,800
C3604	Stepside pickup	133	7,800
C3605	Panel	133	7,800
C3609	Stake bed	133	7,800
C3612	Chassis & cowl (windshield)	133	7,800

Engine & Transmission Suffix Codes

El Camino

G, GB—194ci I-6 1bbl 120hp—Three-speed manual transmission
GF, GG—194ci I-6 1bbl 120hp—Three-speed manual w/PCV
GK, GL, GN—194ci I-6 1bbl 120hp—Three-speed manual w/AC
GM—194ci I-6 1bbl 120hp—Three-speed manual w/AC & PCV
K—194ci I-6 1bbl 120hp—Powerglide automatic
KB—194ci I-6 1bbl 120hp—Powerglide automatic w/PCV
KH—194ci I-6 1bbl 120hp—Powerglide automatic w/AC
KJ—194ci I-6 1bbl 120hp—Powerglide automatic w/AC & PCV
LM—230ci I-6 1bbl 140hp—Three Speed Manual
LL—230ci I-6 1bbl 140hp—Three Speed Manual w/AC
LN—230ci I-6 1bbl 140hp—Three Speed Manual w/AC & PCV
BL—230ci I-6 1bbl 140hp—Powerglide automatic
BN—230ci I-6 1bbl 140hp—Powerglide automatic w/PCV
BM—230ci I-6 1bbl 140hp—Powerglide automatic w/AC
BP—230ci I-6 1bbl 140hp—Powerglide automatic w/AC & PCV
J—283ci V-8 2bbl 195hp—Three Speed manual
JA—283ci V-8 2bbl 195hp—Four-speed manual
JD—283ci V-8 2bbl 195hp—Powerglide automatic
JH—283ci V-8 4bbl 220hp—Three or four-speed manual
JG—283ci V-8 4bbl 220hp—Powerglide automatic
JQ—327ci V-8 4bbl 250hp—Three or four-speed manual
JT—327ci V-8 4bbl 250hp—Three or four-speed manual w/Transistor ignition
SR—327ci V-8 4bbl 250hp—Powerglide automatic
JR—327ci V-8 4bbl 300hp—Three or four-speed manual
SS—327ci V-8 4bbl 300hp—Powerglide automatic
JS—327ci V-8 4bbl NAhp—Four-speed manual, Spec. High Perf.

Corvair 95

V—164ci F-6 2-1bbl 95hp—manual trans
W—164ci F-6 2-1bbl 95hp—Powerglide
VB—164ci F-6 2-1bbl 110hp—manual trans
WB—164ci F-6 2-1bbl 110hp—Powerglide

C-10, C-20 Series

N—230ci I-6 1bbl 140hp—manual trans
ND—230ci I-6 1bbl 140hp—Powerglide
NE—230ci I-6 1bbl 140hp—manual trans
PG—292ci I-6 1bbl 165hp—manual trans
PK—292ci I-6 1bbl 165hp—Powerglide
M—283ci V-8 2bbl 175hp—manual trans
MA—283ci V-8 2bbl 175hp—Powerglide
MX—283ci V-8 2bbl 175hp—Manual
MY—283ci V-8 2bbl 175hp—Powerglide

C-30 Series

NE—230ci I-6 1bbl 140hp—manual trans
PG—292ci I-6 1bbl 165hp—manual trans
M—283ci V-8 2bbl 175hp—manual trans
MX—283ci V-8 2bbl 175hp—manual trans

K-10, K-20 Series

N—230ci I-6 1bbl 140hp—manual trans
NE—230ci I-6 1bbl 140hp—manual trans
PG—292ci I-6 1bbl 165hp—manual trans
M—283ci V-8 2bbl 175hp—manual trans
MX—283ci V-8 2bbl 175hp—manual trans

Transmission Codes

Code	Type	Plant
C	Powerglide	Cleveland (El Camino)
M	3 speed O/D	Muncie
N	4 speed	Muncie
P	4 speed	Muncie (El Camino)
R	4 speed	Saginaw (Corvair)
S	3 speed	Saginaw
T	Powerglide	Toledo
W	3 speed	Warner Gear

Axle Identification

Code	Ratio

El Camino

LA, YA, YG, ZA, ZC, ZE, ZG	3.08:1
LB, LH, YB, YH, ZB, ZF, ZH	3.36:1
ZJ, ZK, ZL, ZM	3.70:1

Corvair

HQ, HR, HS, HT	3.55:1

C-10, K-10 Series

WB, WF	3.07:1
WA, WE, WG WJ	3.73:1
WE, WH,	4.11:1

C-20, K-20 Series

MA, MD	4.11:1
MH, MJ, MK MP, MQ MM, MV	
MX, MZ	4.57:1
MS	5.14:1

C-30 Series

PA, PB PL, PN	5.14:1

Exterior Color Codes

C-10, C-20, C-30 Series

Jet Black	500
Light Green	503
Dark Green	505
Light Blue	507
Dark Blue	508
Turquoise	510
Red	514
Orange	516
Yellow	519
White	521
Gray	522
Off-White	526
Fawn	528
Gray-Green	529

El Camino

Tuxedo Black	900
Meadow Green	905
Bahama Green	908
Silver Blue	912
Daytona Blue	916
Azure Aqua	918
Lagoon Aqua	919
Almond Fawn	920
Ember Red	922
Saddle Tan	932
Ermine White	936
Desert Beige	938
Satin Silver	940
Goldwood Yellow	943
Palomar Red	948

Regular Production Options

El Camino

13380	El Camino 6cyl	$2,267
13480	El Camino 8cyl	2,367
13580	Custom El Camino	2,342
13680	Custom El Camino	2,442

Corvair Series

R1254	Rampside pickup	2,136

C-10 Series

C1403	Chassis and cab	1,893
C1404	Stepside pickup	2,007
C1405	Panel delivery	2,324
C1406	Suburban carryall w/rear doors	2,629
C1416	Suburban carryall w/end-gate	2,662
C1434	Fleetside pickup	2,023
C1503	Chassis & cab	1,931
C1504	Stepside pickup	2,044
C1534	Fleetside pickup	2,060
C1434	Fleetside pickup	2,023
K1403	Chassis and cab, 4WD	2,544
K1404	Stepside pickup 4WD	2,658
K1405	Panel delivery, 4WD	2,975
K1434	Fleetside pickup 4WD	2,674
K1406	Suburban carryall w/rear doors 4WD	3,280
K1416	Suburban carryall w/rear gate 4WD	3,313

C-20 Series

C2503	Chassis & cab	2,078
C2504	Stepside pickup	2,192
C2509	Stake bed	2,283
C2534	Fleetside pickup	2,208
K2503	Chassis & cab 4WD	2,757
K2504	Stepside pickup 4WD	2,869
C2534	Fleetside pickup 4WD	2,885

C-30 Series

C3603	Chassis & cab	2,235
C3604	Stepside pickup	2,370
C3605	Panel	2,791
C3609	Stake bed	2,482

El Camino

Option Number

A01	Soft ray tinted glass
A02	Soft ray tinted glass (windshield only)
A31	Power windows

A41	Four way electric power seat
A49	Custom deluxe seat belts w/retractors
B70	Padded instrument panel
C48	Heater & defroster deletion
C50	Rear window defroster
C60	All season A/C
C65	Custom Deluxe A/C
F40	Special front & rear suspension
G76	3.36:1 rear axle ratio (inc. w/A/C)
G80	Positraction rear axle
J50	Power brakes
J65	Special brakes w/metallic facings
K24	Closed engine positive engine ventilation type B
K02	Temperature controlled radiator fan
K77	55 amp Delcotron generator
K79	42 amp Delcotron generator
K81	62 amp Delcotron generator
L30	250hp Turbo-Fire 328ci V-8
L61	155hp Turbo-Thrift 230ci engine
L77	220hp Turbo-Fire 283ci V-8
M01	Heavy duty clutch
M10	Overdrive transmission
M20	Four speed transmission
M35	Powerglide transmission
N33	7 position Comfortilt steering wheel
N40	Power steering
PO1	Wheel trim cover
P02	Simulated wire wheel covers
T60	Heavy duty 66 plate 70 amp/hr battery
U16	Tachometer
U60	Radio—manual
U63	Radio—push button
V31	Front bumper guard
Z01	Comfort & convenience equipment type A
Z13	Comfort & convenience equipment type B

C-10, C-20, C-30 Series

Option Number

AO9	Laminated glass
A10	Full view rear window
A11	Soft ray tinted glass
A12	Rear window glass
A13	Side door glass
A37	Seat belts, custom deluxe
A54	Full width front seat
A55	Bostrom seat
A57	Auxiliary seat

A59	Supplementary seat	K67	Heavy duty starter motor
A62	Front seat belt (delete)	K71	35 amp low cut-in DC generator
A97	Right door lock	K77	55 amp Delcotron generator
B78	Dispatch box door equipment	K79	42 amp Delcotron generator
B98	Custom side molding	K81	62 amp Delcotron generator
C14	Electric windshield wipers & washer	L05	130 amp Delcotron generator
C40	Direct-air heater & defroster	L25	Engine 292ci I-6
C41	Thrift-air heater & defroster	L26	Engine 230ci I-6
D29	West coast mirror—Jr	L32	Engine, 283ci V-8
D30	West coast mirror—Sr	M01	Heavy duty clutch
D32	Rear view mirror	M16	Warner T89B three speed transmission
E56	Platform and stake rack	M45	Powermatic transmission
E57	Platform body	M55	Powerglide transmission
E80	Pickup box mounting brackets	NO1	Fuel tank—20 gallon
E82	Level box pickup floor	NO2	Fuel tank—30 gallon
E85	Body side door equipment	N12	Exhaust system—single stack
FO3	Heavy duty frame	N13	Exhaust system—dual stack
F49	Heavy duty four wheel drive front axle	N34	Sports-styled walnut grained steering wheel w/plastic ring
F51	Front and rear HD shock absorbers	N40	Power steering
		PO1	Wheel trim cover
F59	Front stabilizer bar	P10	Spare wheel carrier—under frame mounting
F60	Heavy duty front springs		
F76	Front wheel locking hubs	P13	Spare wheel carrier—side mounted
F81	Front springs, HD		
G50	Heavy duty rear springs	T60	Heavy duty battery
G52	7,500lb rear springs	U16	Tachometer
G55	8,750lb rear springs	U60	Radio—manual
G60	Auxiliary rear springs	V01	HD radiator
G80	Limited slip rear axle	V04	Radiator shutters
G81	Positraction differential	V35	Wraparound front bumper
G82	No-Spin rear axle	V37	Custom chrome option
J70	Vacuum brakes	V38	Rear bumper—painted
J71	Full air brakes	V62	Hydraulic jack
J72	Air-hydraulic brakes	V75	Hazard and marker lights
J73	Heavy duty vacuum brakes	V76	Front tow hooks
J80	Vacuum brake reserve tank	Z02	Push button control radio & rear speaker
J81	Vacuum gauge		
J91	Trailer air brake equipment	Z12	Speedometer, driven gear & fitting
K12	Oil filter		
K23	Positive crankcase ventilation	Z50	Frame reinforcements
K24	Closed engine positive engine ventilation type B	Z52	Full depth foam seat
		Z53	Gauges
K28	Fuel filter equipment	Z54	Maximum economy equipment
K37	Engine governor	Z60	Custom equipment
K47	Air cleaner equipment	Z61	Custom appearance option
K48	Oil bath air cleaner	Z62	Custom comfort option
K56	Air compressor equipment	Z72	Vacuum equipment

Facts

The major styling change for 1964 was a redesigned windshield. The front A-pillars leaned back for a more modern appearance. The grille was changed as well, having a tighter mesh, and the headlight buckets were square. The hubcaps were also redesigned.

The interior got a new dash that eliminated the previous pods and was flat. The standard seats were either red or beige, depending on exterior color.

Mechanically, the most notable improvement was the use of self-adjusting brakes on all models. Otherwise, the light trucks were carry-overs.

This was the last year for the Corvair pickup, which came with a more powerful 95hp engine as standard equipment. New was the reintroduction of the El Camino pickup, on the Chevelle station wagon platform. Two versions were available, standard and Custom, in either six- or eight-cylinder variations. Standard on the El Camino was a 194ci inline six, and a 230ci six was optional, as well as 283ci and 327ci V-8s. The standard transmission was a three-speed manual; the Powerglide automatic was optional.

Midyear introductions on the Chevelle and, accordingly, the El Camino, were 250hp and 300hp versions of the 327ci small-block V-8.

No El Camino Super Sports (SSs) were offered in 1964, as the option was limited to the two-door coupe and convertible.

Chapter 21

1965 Trucks

Production

13380 El Camino 6cyl	4,392
13480 El Camino 8cyl	5,935
13580 El Camino Custom 6cyl	2,367
13680 El Camino Custom 8cyl	22,030
Total El Camino	34,724

C-10 Series

C1403 Chassis and cab	4,944
C1404 Stepside pickup	52,899
C1405 Panel delivery	8,228
C1406 Suburban carryall (doors)	4,685
C1416 Suburban carryall (gate)	4,834
C1434 Fleetside pickup	55,300
C1503 Chassis & cab	1,283
C1504 Stepside pickup	27,432
C1534 Fleetside pickup	157,746
K1403 Chassis and cab, 4WD	48
K1404 Stepside pickup 4WD	1,127
K1405 Panel delivery, 4WD	103
K1434 Fleetside pickup 4WD	633
K1406 Suburban carryall (doors) 4WD	444
K1416 Suburban carryall (gate) 4WD	433
K1503 Chassis & cab 4WD	52
K1504 Stepside pickup 4WD	513
K1534 Fleetside pickup 4WD	1,459

C-20 Series

C2503 Chassis & cab	6,267
C2504 Stepside pickup	9,872
C2509 Stake bed	1,505
C2534 Fleetside pickup	46,608
K2503 Chassis & cab 4WD	308
K2504 Stepside pickup 4WD	889
C2534 Fleetside pickup 4WD	1,364
Total 1965 model year (one ton & under)	458,235
Total 1965 model year (all trucks)	552,482

Serial numbers

Description

C1445A100001

C—Chassis type, C—conventional, K—four-wheel drive

14—Series, 14&15—Truck ½ ton, 25—Truck ¾ ton, 36—38—Truck 1-ton

4—Body type, 2—Chassis-cowl, 3—Chassis-cab, 4—Pickup, 5—Panel, 6—Carryall, 9—Platform stake

5—Last digit of model year, 5—1965

A—Assembly Plant Code: A—Atlanta GA, B—Baltimore MD, F—Flint MI, J—Janesville WI, N—Norwood OH, P—Pontiac MI, S—St. Louis MO, T—Tarrytown NY, Z—Fremont CA

100001—Consecutive Sequence Number

Location: On plate attached to left door hinge post.

El Camino Serial Number

136805A100001

1—Chevrolet Division

36—Series, 33—El Camino 6cyl, 34—El Camino 8cyl, 35—Custom El Camino 6cyl, 36—Custom El Camino 8cyl

80—Body style, 2 door sedan pickup
5—Last digit of model year, 5—1965
A—Assembly Plant Code: A—Atlanta GA, B—Baltimore MD, G—Framingham
   MA, K—Kansas City MO, Z—Fremont CA
100001—Consecutive Sequence Number

Location: On plate attached to left door hinge post.

Model Identification
C1004
C—Chassis type, C—conventional, K—four-wheel drive
10—Truck series, 10's—½ ton, 20's—¾ ton, 30's—1 ton
04—Body/truck type, 02—Flat face cowl, 03—Chassis-cab, 04—Stepside
   pickup, 05—Panel, 06—Carryall w/panel rear doors, 09—Stake,
   12—Windshield-cowl, 16—Carryall w/gate, 34—Fleetside pickup

Model, Wheelbase & GVW

Model Number and Description	Wheelbase (in)	GVW
El Camino		
13380 El Camino 6cyl	115	—
13480 El Camino 8cyl	115	—
13580 El Camino Custom 6cyl	115	—
13680 El Camino Custom 8cyl	115	—
C-10 Series		
C1402 Chassis and cowl	115	4,600
C1403 Chassis and cab	115	4,600
C1404 Stepside pickup	115	4,600
C1405 Panel delivery	115	4,600
C1406 Suburban carryall w/rear doors	115	4,600
C1412 Chassis & cowl	115	4,600
C1416 Suburban carryall w/tailgate	115	4,600
C1434 Fleetside pickup	115	5,300
C1503 Chassis & cab	127	5,600
C1504 Stepside pickup	127	4,600
C1534 Fleetside pickup	127	4,600
C1434 Fleetside pickup	127	5,300
K1403 Chassis and cab, 4WD	115	5,300
K1404 Stepside pickup 4WD	115	5,600
K1405 Panel delivery, 4WD	115	5,300
K1434 Fleetside pickup 4WD	115	5,300
K1406 Suburban carryall w/tailgate 4WD	115	5,300
K1416 Suburban carryall w/rear doors 4WD	115	5,300
K1503 Chassis and cab 4WD	127	5,600
K1504 Stepside pickup 4WD	127	5,600
K1534 Fleetside pickup 4WD	127	5,600
C-20 Series		
C2502 Chassis & cowl	127	6,000
C2503 Chassis & cab	127	6,000
C2504 Stepside pickup	127	6,000
C2509 Stake bed	127	6,000
C2534 Fleetside pickup	127	6,000

C2512	Chassis & cowl (windshield)	127	6,000
K2503	Chassis & cab 4WD	127	6,800
K2504	Stepside pickup 4WD	127	6,800
K2534	Fleetside pickup 4WD	127	6,800

C-30 Series

C3602	Chassis & cowl	133	7,800
C3603	Chassis & cab	133	7,800
C3604	Stepside pickup	133	7,800
C3605	Panel	133	7,800
C3609	Stake bed	133	7,800
C3612	Chassis & cowl (windshield)	133	7,800
C3803	Chassis & cab	157	10,000

Engine & Transmission Suffix Codes

El Camino

AA, AC—194ci I-6 1bbl 120hp—Three-speed manual
AG, AH—194ci I-6 1bbl 120hp—Three-speed manual w/AC
AL—194ci I-6 1bbl 120hp—Powerglide automatic
AR—194ci I-6 1bbl 120hp—Powerglide automatic w/AC
CA—230ci I-6 1bbl 140hp—Three-Speed Manual
CB—230ci I-6 1bbl 140hp—Three-speed manual w/AC
CC—230ci I-6 1bbl 140hp—Powerglide automatic
CD—230ci I-6 1bbl 140hp—Powerglide automatic w/AC
DA—283ci V-8 2bbl 195hp—Three Speed manual
DB—283ci V-8 2bbl 195hp—Four-Speed manual
DE—283ci V-8 2bbl 195hp—Powerglide automatic
DG—283ci V-8 4bbl 220hp—Three-speed manual
DH—283ci V-8 4bbl 220hp—Powerglide automatic
EA—327ci V-8 4bbl 250hp—Three or four-speed manual
EE—327ci V-8 4bbl 250hp—Powerglide automatic
EB—327ci V-8 4bbl 300hp—Three or four-speed manual
EF—327ci V-8 4bbl 300hp—Powerglide automatic
ED—327ci V-8 4bbl --hp—Transistor ignition
EC—327ci V-8 4bbl 350hp—Four-speed manual
IX—396ci V-8 4bbl 375hp—Four-speed manual

C-10 Series

TA—230ci I-6 1bbl 140hp—manual trans
TE—230ci I-6 1bbl 140hp—Powerglide
TG—230ci I-6 1bbl 165hp—manual trans
UH—292ci I-6 1bbl 165hp—manual trans
UT—292ci I-6 1bbl 165hp—Powerglide
WA—283ci V-8 2bbl 175hp—manual trans
WE—283ci V-8 2bbl 175hp—Powerglide

C-20 Series

TA—230ci I-6 1bbl 140hp—manual trans
TE—230ci I-6 1bbl 140hp—Powerglide
TG—230ci I-6 1bbl 140hp—manual trans
UH—292ci I-6 1bbl 165hp—manual trans
UT—292ci I-6 1bbl 165hp—Powerglide
WA—283ci V-8 2bbl 175hp—manual trans
WE—283ci V-8 2bbl 175hp—Powerglide

C-30 Series
TG—230ci I-6 1bbl 140hp—manual trans
UH—292ci I-6 1bbl 165hp—manual trans
WA—283ci V-8 2bbl 175hp—manual trans

K-10, K-20 Series
TG—230ci I-6 1bbl 140hp—manual trans
UH—292ci I-6 1bbl 165hp—manual trans
WA—283ci V-8 2bbl 175hp—manual trans

Transmission Codes

Code	Type	Plant
C	Powerglide	Cleveland (El Camino)
M	3 speed OD	Muncie
N	4 speed	Muncie
P	4 speed	Muncie (El Camino)
R	4 speed	Saginaw (Corvair)
S	3 speed	Saginaw
T	Powerglide	Toledo
W	3 speed	Warner Gear

Axle Identification

Code	Ratio

El Camino
GB, GC, GD, GE	2.73:1
CD, CJ, CK, CX	3.07:1
CA, CE, CL, CM	3.08:1
CF, CN, CO, CW	3.31:1
CB, CG, CP, CQ	3.36:1
CH, CR, CS, CU, CV	3.70:1
CT,	3.73:1

C-10, K-10 Series
HB	3.07:1
HA, HC, HE, HF	
HG, HQ	3.73:1
HD, HH,	4.11:1

C-20, K-20 Series
HW	4.11:1
HU, HV, HX, HY, HZ	4.57:1

C-30 Series
IC, ID, IE, IF	5.14:1

Exterior Color Codes

C-10, C-20, C-30 Series
Black	500
Light Green	503
Dark Green	505
Light Blue	507
Dark Blue	508
Turquoise	510
Red	514
Orange	516
Light Yellow	518
Dark Yellow	519
White	521
Gray	522
Off-White	526
Fawn	528

El Camino
Tuxedo Black	A
Ermine White	C
Mist Blue	D
Danube Blue	E
Willow Green	H
Cypress Green	J
Artesian Turquoise	K
Tahitian Turquoise	L
Madeira Maroon	N
Evening Orchid	P
Regal Red	R
Sierra Tan	S
Cameo Beige	V
Glacier Gray	W
Crocus Yellow	Y

Regular Production Options

El Camino
13380 El Camino 6cyl	$2,272
13480 El Camino 8cyl	2,380
13580 Custom El Camino	2,353
13680 Custom El Camino	2,461

C-10 Series

C1403	Chassis and cab	1,894
C1404	Stepside pickup	2,005
C1405	Panel delivery	2,325
C1406	Suburban carryall w/rear doors	2,630
C1416	Suburban carryall w/end-gate	2,665
C1434	Fleetside pickup	2,025
C1503	Chassis & cab	1,930
C1504	Stepside pickup	2,045
C1534	Fleetside pickup	2,060
C1434	Fleetside pickup	2,023
K1403	Chassis and cab 4WD	2,544
K1404	Stepside pickup 4WD	2,660
K1405	Panel delivery 4WD	2,975
K1434	Fleetside pickup 4WD	2,675
K1406	Suburban carryall w/rear doors 4WD	3,280
K1416	Suburban carryall w/rear gate 4WD	3,315

C-20 Series

C2503	Chassis & cab	2,080
C2504	Stepside pickup	2,190
C2509	Stake bed	2,284
C2534	Fleetside pickup	2,210
K2503	Chassis & cab 4WD	2,755
K2504	Stepside pickup 4WD	2,870
C2534	Fleetside pickup 4WD	2,885

C-30 Series

C3603	Chassis & cab	2,235
C3604	Stepside pickup	2,370
C3605	Panel	2,795
C3609	Stake bed	2,483

El Camino

Option Number

A01	Soft ray tinted glass
A02	Soft ray tinted glass (windshield only)
A33	Power windows
A41	Four way electric power seat
A46	Electric control driver's bucket seat
A49	Custom deluxe seat belts w/retractors
A62	Seat belt deletion
B70	Padded instrument panel
C08	Black vinyl roof cover
C48	Heater & defroster deletion
C50	Rear window defroster
C60	All season A/C
C65	Custom Deluxe A/C
F40	Special front & rear suspension

G66	Superlift rear shock absorbers
G67	Superlift rear shock absorbers w/automatic level control
G70	3.70:1 rear axle ratio
G76	3.36:1 rear axle ratio
G80	Positraction rear axle
J50	Power brakes
J65	Special brakes w/metallic facings
K02	Temperature controlled radiator fan
K24	Closed engine positive engine ventilation
K66	Transistorized ignition system
K77	55 amp Delcotron generator
K79	42 amp Delcotron generator
K81	62 amp Delcotron generator
L26	140hp Turbo-Thrift engine
L30	250hp Turbo-Fire engine
L74	300hp Turbo-Thrift engine
L79	350hp Turbo-Fire engine
M01	Heavy duty clutch
M10	Overdrive transmission
M20	Four speed transmission
M35	Powerglide transmission
N10	Dual exhaust
N33	7 position Comfortilt steering wheel
N40	Power steering
PO1	4 brightmetal wheel covers
P02	Simulated wire wheel covers
P19	Spare wheel lock
T60	Heavy duty 66 plate 70 amp/hr battery
U03	Tri-Volume horn
U16	Tachometer
U60	Radio—manual control
U63	Radio—push button control
U69	Radio AM/FM push button control
V01	HD radiator
V31	Front bumper guard
V32	Rear bumper guard
Z01	Comfort & convenience equipment type A
Z13	Comfort & convenience equipment type B

C-10, C-20, C-30 Series

Option Number

AO9	Laminated glass
A10	Full view rear window
A11	Soft ray tinted glass
A12	Rear window glass
A13	Side door glass
A37	Seat belts, custom deluxe

A54	Full width front seat
A55	Bostrom seat
A57	Auxiliary seat
A59	Supplementary seat
A62	Front seat belt (delete)
A97	Right door lock
B78	Dispatch box door equipment
B98	Custom side molding
C14	Electric windshield wipers & washer
C40	Direct-air heater & defroster
C41	Thrift-air heater & defroster
C60	Air conditioning
D29	West coast mirror—Jr
D30	West coast mirror—Sr
D32	Rear view mirror
E56	Platform and stake rack
E57	Platform body
E80	Pickup box mounting brackets
E82	Level box pickup floor
E85	Body side door equipment
FO3	Heavy duty frame
F49	Heavy duty four wheel drive front axle
F51	Front and rear HD shock absorbers
F57	9000lb front axle
F58	11,000lb front axle
F59	Front stabilizer bar
F60	Heavy duty front springs
F76	Front wheel locking hubs
F81	Front springs, HD
G50	Heavy duty rear springs
G52	7,500lb rear springs
G55	8,750lb rear springs
G60	Auxiliary rear springs
G80	Limited slip rear axle
G81	Positraction differential
G82	No-Spin rear axle
J70	Vacuum brakes
J71	Full air brakes
J72	Air-hydraulic brakes
J73	Heavy duty vacuum brakes
J80	Vacuum brake reserve tank
J81	Vacuum gauge
J91	Trailer air brake equipment
K12	Oil filter
K23	Positive crankcase ventilation
K24	Closed engine positive engine ventilation type B
K28	Fuel filter equipment

K37	Engine governor
K47	Air cleaner equipment
K48	Oil bath air cleaner
K56	Air compressor equipment
K67	Heavy duty starter motor
K71	35 amp low cut-in DC generator
K77	55 amp Delcotron generator
K79	42 amp Delcotron generator
K81	62 amp Delcotron generator
L05	130 amp Delcotron generator
L25	Engine 292ci I-6
L26	Engine 230ci I-6
M01	Heavy duty clutch
M16	Warner T89B three speed transmission
M55	Powerglide transmission
NO1	Fuel tank—20 gallon
NO2	Fuel tank—30 gallon
N34	Sports-styled walnut grained steering wheel w/plastic ring
N40	Power steering
PO1	Wheel trim cover
P10	Spare wheel carrier—under frame mounting
P13	Spare wheel carrier—side mounted
T60	Heavy duty battery
U16	Tachometer
U60	Radio—manual
V01	HD radiator
V04	Radiator shutters
V05	HD cooling
V35	Wraparound front bumper
V37	Custom chrome option
V38	Rear bumper—painted
V62	Hydraulic jack
V75	Hazard and marker lights
V76	Front tow hooks
Z02	Push button control radio & rear speaker
Z12	Speedometer, driven gear & fitting
Z50	Frame reinforcements
Z52	Full depth foam seat
Z53	Gauges
Z54	Maximum economy equipment
Z60	Custom equipment
Z61	Custom appearance option
Z62	Custom comfort option
Z72	Vacuum equipment
Z81	Camper special equipment

Facts

The side emblems were relocated to the upper cowl on the 1965 trucks. They were rectangular in shape, enclosing a bow-tie emblem ornament, and had series numerals next to them. Otherwise, the 1965 models were unchanged.

The Z81 Camper Special equipment option was made available. It included the heavy-duty components necessary for the installation of a camper, such as suspension and chassis.

Factory-installed air conditioning was optional for the first time.

Engine and transmission choices remained unchanged from those offered in 1964.

The El Camino was restyled along with the rest of the Chevelle line. The L79 350hp version of the 327ci V-8 was newly optional. It was available only with a four-speed manual transmission.

Although the 396ci big-block was available under the Z16 option package, no examples of it were installed in the El Camino.

Chapter 22

1966 Trucks

Production

13380	El Camino 6cyl	3,424
13480	El Camino 8cyl	5,897
13580	El Camino Custom 6cyl	1,461
13680	El Camino Custom 8cyl	24,337
	Total El Camino	35,119

C-10 Series

C1403	Chassis and cab	3,030
C1404	Stepside pickup	59,947
C1405	Panel delivery	8,344
C1406	Suburban carryall w/rear doors	6,717
C1416	Suburban carryall w/end-gate	5,334
C1434	Fleetside pickup	57,386
C1503	Chassis and cab	1,155
C1504	Stepside pickup	26,456
C1534	Fleetside pickup	178,752
K1403	Chassis and cab 4WD	40
K1404	Stepside pickup 4WD	1,123
K1405	Panel delivery 4WD	170
K1434	Fleetside pickup 4WD	678
K1406	Suburban carryall w/rear doors 4WD	530
K1416	Suburban carryall w/rear gate 4WD	418

C-20 Series

C2503	Chassis & cab	6,520
C2504	Stepside pickup	9,905
C2509	Stake bed	1,499
C2534	Fleetside pickup	55,855
K2503	Chassis & cab 4WD	431
K2504	Stepside pickup 4WD	923
C2534	Fleetside pickup 4WD	1,796

C-30 Series

C3603	Chassis & cab	11,852
C3604	Cab and chassis	3,646
C3605	Panel	3,560
C3609	Stake bed	3,651

Serial numbers

Description

C1445A100001

C—Chassis type, C—conventional, K—four-wheel drive

14—Series, 14&15—Truck ½ ton, 25—Truck ¾ ton, 36&38—Truck 1-ton

4—Body type, 2—Chassis-cowl, 3—Chassis-cab, 4—Pickup, 5—Panel, 6—Carryall, 9—Platform stake

6—Last digit of model year, 6—1966

A—Assembly Plant Code: A—Atlanta GA, B—Baltimore MD, F—Flint MI, J—Janesville WI, N—Norwood OH, P—Pontiac MI, S—St. Louis MO, T—Tarrytown NY, Z—Fremont CA

100001—Consecutive Sequence Number

Location: On plate attached to left door hinge post.

El Camino Serial Number

136806A100001

1—Chevrolet Division

36—Series, 33—El Camino 6cyl, 34—El Camino 8cyl, 35—Custom El Camino 6cyl, 36—Custom El Camino 8cyl

80—Body style, 2 door sedan pickup

6—Last digit of model year, 6—1966
A—Assembly Plant Code: A—Atlanta GA, B—Baltimore MD, F—Flint MI,
   G—Framingham MA, K—Kansas City MO, Z—Fremont CA
100001—Consecutive Sequence Number

Location: On plate attached to left door hinge post.

Model Identification
C1004

C—Chassis type, C—conventional, K—four-wheel drive
10—Truck series, 10's—½ ton, 20's—¾ ton, 30's—1 ton
04—Body/truck type, 02—Flat face cowl, 03—Chassis-cab, 04—Stepside
   pickup, 05—Panel, 06—Carryall w/panel rear doors, 09—Stake,
   12—Windshield-cowl, 16—Carryall w/gate, 34—Fleetside pickup

Model, Wheelbase & GVW

Model Number and Description	Wheelbase (in)	GVW
El Camino		
13380 El Camino 6cyl	115	—
13480 El Camino 8cyl	115	—
13580 El Camino Custom 6cyl	115	—
13680 El Camino Custom 8cyl	115	—
C-10 Series		
C1402 Chassis and cowl	115	4,600
C1403 Chassis and cab	115	4,600
C1404 Stepside pickup	115	4,600
C1405 Panel delivery	115	4,600
C1406 Suburban carryall w/rear doors	115	4,600
C1412 Chassis & cowl	115	4,600
C1416 Suburban carryall w/tailgate	115	4,600
C1434 Fleetside pickup	115	5,300
C1503 Chassis & cab	127	5,600
C1504 Stepside pickup	127	4,600
C1534 Fleetside pickup	127	4,600
C1434 Fleetside pickup	127	5,300
K1403 Chassis and cab 4WD	115	5,300
K1404 Stepside pickup 4WD	115	5,600
K1405 Panel delivery 4WD	115	5,300
K1434 Fleetside pickup 4WD	115	5,300
K1406 Suburban carryall w/tailgate 4WD	115	5,300
K1416 Suburban carryall w/rear doors 4WD	115	5,300
K1503 Chassis and cab 4WD	127	5,600
K1504 Stepside pickup 4WD	127	5,600
K1534 Fleetside pickup 4WD	127	5,600
C-20 Series		
C2502 Chassis & cowl	127	6,000
C2503 Chassis & cab	127	6,000
C2504 Stepside pickup	127	6,000
C2509 Stake bed	127	6,000
C2534 Fleetside pickup	127	6,000
C2512 Chassis & cowl (windshield)	127	6,000

K2503	Chassis & cab 4WD	127	6,800
K2504	Stepside pickup 4WD	127	6,800
K2534	Fleetside pickup 4WD	127	6,800

C-30 Series

C3602	Chassis & cowl	133	7,800
C3603	Chassis & cab	133	7,800
C3604	Stepside pickup	133	7,800
C3605	Panel	133	7,800
C3609	Stake bed	133	7,800
C3612	Chassis & cowl (windshield)	133	7,800
C3803	Chassis & cab	157	10,000

Engine & Transmission Suffix Codes

El Camino

AA, AC—194ci I-6 1bbl 120hp—Three-speed manual
AG, AH—194ci I-6 1bbl 120hp—Three-speed manual w/AC
AL—194ci I-6 1bbl 120hp—Powerglide automatic
AR—194ci I-6 1bbl 120hp—Powerglide automatic w/AC
CA—230ci I-6 1bbl 140hp—Three-Speed Manual
CB—230ci I-6 1bbl 140hp—Three-Speed manual w/AC
CC—230ci I-6 1bbl 140hp—Powerglide automatic
CD—230ci I-6 1bbl 140hp—Powerglide automatic w/AC
BN, BO—230ci I-6 1bbl 140hp—Three-speed manual w/AC & w/AIR
BL, BM—230ci I-6 1bbl 140hp—Powerglide w/AC & w/AIR
DA—283ci V-8 2bbl 195hp—Three Speed manual
DB—283ci V-8 2bbl 195hp—Four-Speed manual
DE—283ci V-8 2bbl 195hp—Powerglide automatic
DI—283ci V-8 2bbl 195hp—Three-speed manual
DJ—283ci V-8 2bbl 195hp—Powerglide w/AIR
DG—283ci V-8 4bbl 220hp—Three-speed manual
DH—283ci V-8 4bbl 220hp—Powerglide automatic
EA—327ci V-8 4bbl 250hp—Three or four-speed manual
EC—327ci V-8 4bbl 250hp—Powerglide automatic
EB—327ci V-8 4bbl 250hp—Three or four-speed manual
EE—327ci V-8 4bbl 250hp—Powerglide automatic
ED—396ci V-8 4bbl 325hp—Three or four-speed manual
EH—396ci V-8 4bbl 325hp—Three or four-speed w/AIR
EK—396ci V-8 4bbl 325hp—Powerglide automatic
EM—396ci V-8 4bbl 325hp—Powerglide automatic w/AIR
EF—396ci V-8 4bbl 360hp—Three or four-speed manual
EJ—396ci V-8 4bbl 360hp—Three or four-speed w/AIR
EL—396ci V-8 4bbl 360hp—Powerglide automatic
EN—396ci V-8 4bbl 360hp—Powerglide automatic w/AIR
EG—396ci V-8 4bbl 375hp—Four-speed manual

C-10 Series

TA—250ci I-6 1bbl 155hp—manual trans
TE—250ci I-6 1bbl 155hp—Powerglide
TF—250ci I-6 1bbl 155hp—manual trans
SV—250ci I-6 1bbl 155hp—manual trans w/AIR
SW—250ci I-6 1bbl 155hp—Powerglide
SX—250ci I-6 1bbl 155hp—manual trans
UH—292ci I-6 1bbl 170hp—manual trans
UT—292ci I-6 1bbl 170hp—Powerglide
VQ—292ci I-6 1bbl 170hp—manual trans

VR—292ci I-6 1bbl 170hp—Powerglide
WA—283ci V-8 2bbl 175hp—manual trans
WE—283ci V-8 2bbl 175hp—Powerglide
WF—283ci V-8 2bbl 175hp—manual trans w/AIR
WH—283ci V-8 2bbl 175hp—Powerglide w/AIR
YS—327ci V-8 4bbl 220hp—manual trans
YR—327ci V-8 4bbl 220hp—Powerglide
YC—327ci V-8 4bbl 220hp—manual trans w/AIR
YD—327ci V-8 4bbl 220hp—Powerglide w/AIR

C-20 Series
TA—230ci I-6 1bbl 155hp—manual trans
TE—230ci I-6 1bbl 155hp—automatic trans
TF—250ci I-6 1bbl 155hp—manual trans
VH—292ci I-6 1bbl 170hp—manual trans
UT—292ci I-6 1bbl 170hp—automatic trans
WA—283ci V-8 2bbl 175hp—manual trans
WE—283ci V-8 2bbl 175hp—Powerglide
YS—327ci V-8 4bbl 220hp—manual trans
YR—327ci V-8 4bbl 220hp—Powerglide
YH—327ci V-8 4bbl 220hp—Hydramatic

C-30 Series
TE—230ci I-6 1bbl 155hp—automatic trans
TF—250ci I-6 1bbl 155hp—manual trans
UH—292ci I-6 1bbl 165hp—manual trans
WA—283ci V-8 2bbl 175hp—manual trans
YS—327ci V-8 4bbl 220hp—manual trans
YH—327ci V-8 4bbl 220hp—Hydramatic

K-10 Series
TA—230ci I-6 1bbl 155hp—manual trans
SV—250ci I-6 1bbl 155hp—manual trans w/AIR
SX—250ci I-6 1bbl 155hp—manual trans
UH—292ci I-6 1bbl 170hp—manual trans
VQ—292ci I-6 1bbl 170hp—manual trans
WA—283ci V-8 2bbl 175hp—manual trans
WF—283ci V-8 2bbl 175hp—manual trans

K-20 Series
TA—230ci I-6 1bbl 155hp—manual trans
UH—292ci I-6 1bbl 170hp—manual trans
WA—283ci V-8 2bbl 175hp—manual trans

Transmission Codes

Code	Type	Plant
C	Powerglide	Cleveland (El Camino)
CA	Hydra-Matic	—
M	3 speed OD	Muncie
N	4 speed	Muncie
P	4 speed	Muncie (El Camino)
R	4 speed	Saginaw
S	3 speed	Saginaw
T	Powerglide	Toledo
W	3/4 speed	Warner Gear

Axle Identification

Code	Ratio
El Camino	
KC, KD, KE, KI	2.73:1
CD, CJ, CK, CX	3.07:1

118

CA, CE, CL, CM	3.08:1
CF, CN, CO, CW	3.31:1
CB, CG, CP, CQ	3.36:1
CH, CR, CS, CV	3.70:1
CT, CU	3.73:1
KK, KL	4.10:1
KM, KN, KO, KP	4.56:1

C-10, K-10 Series

HB	3.07:1
HA, HC, HE, HF	
HG, HQ	3.73:1
HD, HH,	4.11:1

C-20, K-20 Series

HW, JU	4.11:1
HU, HV, HX, HY, HZ	
JS, JT	4.57:1

C-30 Series

LW, JN	4.57:1
IC, ID, IE, IF	
LU, LV	5.14:1

Exterior Color Codes

C-10, C-20, C-30 Series

Black	500
Light Green	503
Dark Green	505
Light Blue	507
Dark Blue	508
Turquoise	510
Dark Aqua	511
Red	514
Orange	516
Dark Yellow	519
White	521
Gray	522
Silver	523
Saddle	525
Ivory	526

El Camino

Tuxedo Black	A
Ermine White	C
Mist Blue	D
Danube Blue	E
Marina Blue	F
Willow Green	H
Artesian Turquoise	K
Tropic Turquoise	L
Aztec Bronze	M
Madeira Maroon	N
Regal Red	R
Sandalwood Tan	T
Cameo Beige	V
Chateau Slate	W
Lemonwood Yellow	Y

Regular Production Options

El Camino

13380	El Camino 6cyl	$2,318
13480	El Camino 8cyl	2,426
13580	Custom El Camino	2,396
13680	Custom El Camino	2,504

C-10 Series

C1403	Chassis and cab	1,927
C1404	Stepside pickup	2,050
C1405	Panel delivery	2,361
C1406	Suburban carryall w/rear doors	2,598
C1416	Suburban carryall w/end-gate	2,629
C1434	Fleetside pickup	2,718
C1504	Stepside pickup	2,087
C1534	Fleetside pickup	2,104
C1434	Fleetside pickup	2,066
K1403	Chassis and cab 4WD	2,579
K1404	Stepside pickup 4WD	2,702
K1405	Panel delivery 4WD	2,993
K1434	Fleetside pickup 4WD	2,718
K1406	Suburban carryall w/rear doors 4WD	3,250
K1416	Suburban carryall w/rear gate 4WD	3,281

C-20 Series

C2503	Chassis & cab	2,112
C2504	Stepside pickup	2,236
C2509	Stake bed	2,358
C2534	Fleetside pickup	2,252
K2503	Chassis & cab 4WD	2,789
K2504	Stepside pickup 4WD	2,913
C2534	Fleetside pickup 4WD	2,929

C-30 Series

C3603	Chassis & cab	2,269
C3605	Panel	2,832
C3609	Stake bed	2,527

El Camino

Option Number

A01	Soft ray tinted glass
A02	Soft ray tinted glass (windshield only)
A31	Power windows
A39	Custom deluxe seat belts w/retractors

A41	Four way electric power seat
A46	6 way power front seat
A49	Custom deluxe seat belts w/retractors
A51	Strato-Bucket seats
A81	Strato-Ease headrests
A82	Strato-Ease headrests
B90	Side window moldings
C08	Black vinyl roof cover
C48	Heater & defroster deletion
C50	Rear window defroster
C51	Rear window air deflector
C60	Four season A/C
F40	Special front & rear suspension
G66	Superlift rear shock absorbers
G76	3.36:1 rear axle ratio
G80	Positraction rear axle
G94	3.31:1 rear axle ratio
G96	3.55:1 rear axle ratio
H01	3.07:1 rear axle ratio
J50	Power brakes
J65	Special brakes w/metallic facings
K02	Temperature controlled radiator fan
K19	AIR equipment
K24	Closed engine positive engine ventilation
K66	Transistorized ignition system
K76	61 amp Delcotron generator
K79	42 amp Delcotron generator
K81	62 amp Delcotron generator
L26	140hp Turbo-Thrift six cylinder engine
L30	275hp Turbo-Fire 327ci engine
L34	360hp Turbo-Jet 396ci engine
L77	220hp Turbo-Fire 283ci engine
L78	375hp Turbo-Fire 396ci engine
M01	Heavy duty clutch
M10	Overdrive transmission
M20	Four speed transmission
M21	Four speed close ratio transmission
M35	Powerglide transmission
M55	Transmission oil cooler
N10	Dual exhaust
N33	7 position Comfortilt steering wheel
N34	Sports styled steering wheel
N40	Power steering
N96	Mag style wheel covers
PO1	4 brightmetal wheel covers
P02	Simulated wire wheel covers
P19	Spare wheel lock
T60	Heavy duty 66 plate 70 amp/hr battery
U03	Tri-Volume horn
U14	Special instrumentation

U16	Tachometer
U63	Radio—push button control
U69	Radio AM/FM push button control
V01	HD radiator
V31	Front bumper guard
V32	Rear bumper guard
V74	Traffic hazard warning switch

C-10, C-20, C-30 Series

Option Number

AO9	Laminated glass
A10	Full view rear window
A11	Soft ray tinted glass
A12	Rear window glass
A13	Side door glass
A37	Seat belts, custom deluxe
A54	Full width front seat
A55	Bostrom seat
A57	Auxiliary seat
A59	Supplementary seat
A62	Front seat belt (delete)
A97	Right door lock
B78	Dispatch box door equipment
B98	Custom side molding
C14	Electric windshield wipers & washer
C40	Direct-air heater & defroster
C41	Thrift-air heater & defroster
C60	Air conditioning
D29	West coast mirror—Jr
D30	West coast mirror—Sr
D32	Rear view mirror
E56	Platform and stake rack
E57	Platform body
E80	Pickup box mounting brackets
E82	Level box pickup floor
E85	Body side door equipment
FO3	Heavy duty frame
F49	Heavy duty four wheel drive front axle
F51	Front and rear HD shock absorbers
F57	9000lb front axle
F59	Front stabilizer bar
F60	Heavy duty front springs
F76	Front wheel locking hubs
F81	Front springs, HD
G50	Heavy duty rear springs
G52	7,500lb rear springs
G55	8,750lb rear springs
G60	Auxilliary rear springs
G80	Limited slip rear axle
G81	Positraction differential
G82	No-Spin rear axle
J70	Vacuum brakes

J71	Full air brakes	N40	Power steering
J72	Air-hydraulic brakes	PO1	Wheel trim cover
J73	Heavy duty vacuum brakes	P10	Spare wheel carrier—under frame mounting
J80	Vacuum brake reserve tank		
J81	Vacuum gauge	P13	Spare wheel carrier—side mounted
J91	Trailer air brake equipment		
K12	Oil filter	T60	Heavy duty battery
K19	AIR equipment	U16	Tachometer
K23	Positive crankcase ventilation	U60	Radio—manual
K24	Closed engine positive engine ventilation type B	V01	HD radiator
		V04	Radiator shutters
K28	Fuel filter equipment	V05	HD cooling
K37	Engine governor	V35	Wraparound front bumper
K47	Air cleaner equipment	V37	Custom chrome option
K48	Oil bath air cleaner	V38	Rear bumper—painted
K56	Air compressor equipment	V62	Hydraulic jack
K67	Heavy duty starter motor	V75	Hazard and marker lights
K71	35 amp low cut-in DC generator	V76	Front tow hooks
K77	55 amp Delcotron generator	Z02	Push button control radio & rear speaker
K79	42 amp Delcotron generator		
K81	62 amp Delcotron generator	Z12	Speedometer, driven gear & fitting
L05	130 amp Delcotron generator		
M01	Heavy duty clutch	Z50	Frame reinforcements
M16	Warner T89B three speed transmission	Z52	Full depth foam seat
		Z53	Gauges
M45	Powermatic transmission	Z54	Maximum economy equipment
M55	Powerglide transmission	Z60	Custom equipment
NO1	Fuel tank—20 gallon	Z61	Custom appearance option
NO2	Fuel tank—30 gallon	Z62	Custom comfort option
N12	Exhaust system—single stack	Z72	Vacuum equipment
N13	Exhaust system—dual stack	Z81	Camper special equipment
N34	Sports-styled walnut grained steering wheel w/plastic ring		

Facts

Visually, the 1965 and 1966 trucks were almost indistinguishable. You can tell the difference by looking at the side emblems. For 1966, these incorporated the Chevy bow-tie within a rectangular shape and were placed above the series numerals.

The base engine was a larger 250ci inline six rated at 150hp. The power rating of the optional 292ci six went up 5hp to 170hp, and the 283ci V-8 remained unchanged at 175hp. An additional V-8 was made available: the 327ci small-block V-8 rated at 220hp.

The Saginaw three-speed manual overdrive transmission was fully synchronized.

Additional standard features included seatbelts, an outside rearview mirror, two-speed electric windshield wipers and washers, and back-up lamps. These were federally mandated.

The Custom Camper option on the C-20s included larger, 7.5x16 tires; West Coast mirrors; a tinted windshield; heavy-duty shocks; auxiliary springs; a front stabilizer bar; a deluxe heater and defroster; an AM radio; and dual sun visors. The Camper option also got the Custom Exterior Package with its chrome bumper, grille, and moldings and brightmetal interior trim. A total 9,144 trucks were so equipped.

The 10 millionth Chevrolet truck was built This year.

This was the last year for the V-8—era trucks.

Although the El Camino could not be equipped with the SS Option Package, most of the individual SS components could be had separately. The 396ci big-block V-8 was available in L35 form, where it was rated at 325hp. It came with a hydraulic camshaft and a Rochester Quadrajet carburetor. A total 1,865 El Caminos were so equipped.

Chapter 23

1967 Trucks

Production

13380 El Camino 6cyl	2,963
13480 El Camino 8cyl	5,387
13580 El Camino Custom 6cyl	1,098
13680 El Camino Custom 8cyl	25,382
Total El Camino	34,830

C-10 Series

		K Series
CE10702 Chassis and cowl		
CE10703 Chassis and cab	2,790	39
CE10704 Stepside pickup	45,606	1,229
CE10734 Fleetside pickup	43,940	1,046
CE10904 Stepside pickup	19,969	500
CE10905 Panel delivery	3,827	30
CE10906 Suburban w/rear doors	5,206	509
CE10934 Fleetside pickup	165,973	2,715

C-20 Series

CE20903 Chassis and cab	6,320	498
CE20904 Stepside pickup	7,859	872
CE20905 Panel delivery	940	8
CE20906 Suburban w/rear doors	709	120
CE20909 Stake bed	1,415	—
CE20934 Fleetside pickup	50,413	2,773

C-30 Series

CE31001 Chassis and cowl	366
CS31003 Chassis & cab	11,304
CS31004 Stepside pickup	4,026
CS31009 Stake bed	3,286
CS31012 Cowl and windshield	8
CS31043 Chassis & cab	4,488
1967 Model Year Production	526,776

Serial numbers

Description

CE107047A100001
C—Chassis type, C—conventional, K—four-wheel drive
E—Engine: E—V-8, S—V-6
1—GVW range: 1—3600-5600lb, 2—5500-8100lb, 3—6700-10,000lb
07—Cab to axle dimension: 07—42-47″, 09—54-59″, 10—60-65″, 14—84-89″
04—Body type, 02—Cowl, 03—Cab, 04—Stepside pickup, 05—Panel, 06—
   Suburban (panel rear doors), 09—Platform stake, 12—Windshield cowl,

13—Cab (full air brakes), 16—Suburban (tail & liftgate), 34—Fleetside pickup
7—Last digit of model year, 7—1967
A—Assembly Plant Code: A—Atlanta GA, B—Baltimore MD, F—Flint MI, J—Janesville WI, K—Kansas City MO, S—St. Louis MO, T—Tarrytown NY, Z—Fremont CA, 1—Oshawa Canada
100001—Consecutive Sequence Number

Location: On plate attached to left door hinge post. On cowl models, plate is attahed to left cowl inner panel.

El Camino Serial Number

136806A100001
1—Chevrolet Division
36—Series, 33—El Camino 6cyl, 34—El Camino 8cyl, 35—Custom El Camino 6cyl, 36—Custom El Camino 8cyl
80—Body style, 2 door sedan pickup
7—Last digit of model year, 7—1967
A—Assembly Plant Code: A—Atlanta GA, B—Baltimore MD, G—Framingham MA, K—Kansas City MO, Z—Fremont CA
100001—Consecutive Sequence Number

Location: On plate attached to left door hinge post.

Model & Wheelbase

Model Number and Description

Model Number and Description	Wheelbase (in)
El Camino	
13380 El Camino 6cyl	115
13480 El Camino 8cyl	115
13580 El Camino Deluxe 6cyl	115
13680 El Camino Deluxe 8cyl	115
C-10 Series	
CE10702 Chassis and cowl	115
CE10703 Chassis and cab	115
CE10704 Stepside pickup	115
CE10712 Windshield Cowl	115
CE10734 Fleetside pickup	115
CE10903 Chassis & cab	127
CE10904 Stepside pickup	127
CE10905 Panel delivery	127
CE10906 Suburban w/rear doors	127
CE10916 Suburban w/tailgate	127
CE10934 Fleetside pickup	127
C-20 Series	
CE20902 Chassis and cowl	127
CE20903 Chassis and cab	127
CE20904 Stepside pickup	127
CE20905 Panel delivery	127
CE20906 Suburban w/rear doors	127
CE20909 Stake bed	127

CE20912	Windshield Cowl	127
CE20916	Suburban w/tailgate	127
CE20934	Fleetside pickup	133

C-30 Series

CE31002	Chassis & cowl	133
CE31003	Chassis & cab	133
CE31004	Stepside pickup	133
CE31009	Stake bed	133
CE31012	Chassis & cowl (windshield)	133
CE31034	Stepside pickup	133
CE31403	Chassis & cab	157

K-10 Series

KE10703	Chassis and cab	115
KE10704	Stepside pickup	115
KE10905	Panel deliery	115
KE10906	Suburban w/rear doors	127
KE10916	Suburban w/tailgate	127
KE10734	Fleetside pickup	115
KE10903	Chassis & cab	127
KE10904	Stepside pickup	127
KE10934	Fleetside pickup	127

K-20 Series

KE20903	Chassis and cab	127
KE20904	Stepside pickup	127
KE20905	Panel delivery	127
KE20906	Suburban w/rear doors	127
KE20916	Suburban w/tailgate	127
KE20934	Fleetside pickup	127

Engine & Transmission Suffix Codes

El Camino

CA, CB—230ci I-6 1bbl 140hp—Three-Speed Manual
CC, CD—230ci I-6 1bbl 140hp—Three-Speed manual w/AC
BN, BO—230ci I-6 1bbl 140hp—Three-speed manual w/AC & w/AIR
BL, BM—230ci I-6 1bbl 140hp—Powerglide w/AC & w/AIR
CM, CN—250ci I-6 1bbl 155hp—Manual trans
CQ, CR—250ci I-6 1bbl 155hp—Powerglide
CO, CP—250ci I-6 1bbl 155hp—Manual trans
CS, CT—250ci I-6 1bbl 155hp—Powerglide
DA—283ci V-8 2bbl 195hp—Three Speed manual
DB—283ci V-8 2bbl 195hp—Four-speed manual
DE—283ci V-8 2bbl 195hp—Powerglide automatic
DI—283ci V-8 2bbl 195hp—Three-speed manual
DJ—283ci V-8 2bbl 195hp—Powerglide w/AIR
DG—283ci V-8 2bbl 195hp—Three-speed manual
DK—283ci V-8 2bbl 195hp—Four-speed manual w/AIR
EA—327ci V-8 4bbl 275hp—Manual trans
EB—327ci V-8 4bbl 275hp—Manual w/AIR
EC—327ci V-8 4bbl 275hp—Powerglide w/AIR
EE—327ci V-8 4bbl 275hp—Powerglide
EQ—327ci V-8 4bbl 275hp—Manual trans w/HD clutch
EP—327ci V-8 4bbl 325hp—Manual trans

ES—327ci V-8 4bbl 325hp—Manual trans w/HD clutch
ER—327ci V-8 4bbl 325hp—Manual trans w/AIR
ED—396ci V-8 4bbl 325hp—Three or four-speed manual
EH—396ci V-8 4bbl 325hp—Three or four-speed manual w/AIR
EK—396ci V-8 4bbl 325hp—Powerglide automatic
EM—396ci V-8 4bbl 325hp—Powerglide automatic w/AIR
ET—396ci V-8 4bbl 325hp—Turbo-Hydramatic automatic
EV—396ci V-8 4bbl 325hp—Turbo-Hydramatic automatic w/AIR
EF—396ci V-8 4bbl 350hp—Three or four-speed manual
EJ—396ci V-8 4bbl 350hp—Three or four-speed w/AIR
EL—396ci V-8 4bbl 350hp—Powerglide automatic
EN—396ci V-8 4bbl 350hp—Powerglide automatic w/AIR
EU—396ci V-8 4bbl 350hp—Turbo-Hydramatic Automatic
EW—396ci V-8 4bbl 350hp—Turbo-Hydramatic Automatic w/AIR
EG—396ci V-8 4bbl 375hp—Four-speed manual
EX—396ci V-8 4bbl 375hp—Four-speed manual w/AIR

C-10 Series

TA—250ci I-6 1bbl 155hp—manual trans
TC—250ci I-6 1bbl 155hp—manual trans
TD—250ci I-6 1bbl 155hp—four-speed manual
TE—250ci I-6 1bbl 155hp—automatic
TF—250ci I-6 1bbl 155hp—four-speed manual
TS—292ci I-6 1bbl 170hp—Turbo-Hydramatic w/AIR
SV—250ci I-6 1bbl 155hp—manual trans w/AIR
SW—250ci I-6 1bbl 155hp—automatic w/AIR
SX—250ci I-6 1bbl 155hp—four-speed manual/HD
UH—292ci I-6 1bbl 170hp—manual trans
UR—292ci I-6 1bbl 170hp—manual trans
UQ—292ci I-6 1bbl 170hp—Turbo-Hydramatic
UT—292ci I-6 1bbl 170hp—Powerglide
VQ—292ci I-6 1bbl 170hp—manual trans w/AIR
VR—292ci I-6 1bbl 170hp—Powerglide w/AIR
WA—283ci V-8 2bbl 175hp—manual trans
WC—283ci V-8 2bbl 175hp—Turbo-Hydramatic
WE—283ci V-8 2bbl 175hp—Powerglide
WF—283ci V-8 2bbl 175hp—manual trans w/AIR
WG—283ci V-8 2bbl 175hp—manual trans
WH—283ci V-8 2bbl 175hp—Powerglide w/AIR
WR—283ci V-8 2bbl 175hp—Turbo-Hydramatic w/AIR
YS—327ci V-8 4bbl 220hp—manual trans
YR—327ci V-8 4bbl 220hp—Powerglide
YC—327ci V-8 4bbl 220hp—manual trans w/AIR
YD—327ci V-8 4bbl 220hp—Powerglide w/AIR
YJ—327ci V-8 4bbl 220hp—Turbo-Hydramatic w/AIR
YH—327ci V-8 4bbl 220hp—Powerglide w/AIR

C-30 Series

TE—230ci I-6 1bbl 155hp—automatic trans
TF—250ci I-6 1bbl 155hp—four-speed manual
UH—292ci I-6 1bbl 170hp—manual trans
UR—292ci I-6 1bbl 170hp—manual trans
WA—283ci V-8 2bbl 175hp—manual trans
WG—283ci V-8 2bbl 175hp—manual trans
YS—327ci V-8 4bbl 220hp—manual trans

K-10, K-20 Series

TA—250ci I-6 1bbl 155hp—manual trans

TC—250ci I-6 1bbl 155hp—manual trans
TO—250ci I-6 1bbl 155hp—four-speed manual
SV—250ci I-6 1bbl 155hp—manual trans w/AIR
SX—250ci I-6 1bbl 155hp—four-speed manual
UH—292ci I-6 1bbl 170hp—manual trans
VQ—292ci I-6 1bbl 170hp—manual trans
VR—292ci I-6 1bbl 170hp—manual trans
WB—283ci V-8 2bbl 175hp—manual trans
WO—283ci V-8 2bbl 175hp—manual trans
WI—283ci V-8 2bbl 175hp—manual trans
YX—327ci V-8 4bbl 220hp—manual trans
YM—327ci V-8 4bbl 220hp—manual trans

Transmission Codes

Code	Type	Plant
C	Powerglide	Cleveland (El Camino)
CA	Hydra-Matic	—
E	Powerglide	McKinnon
K	3 speed	McKinnon
M	3 speed OD	Muncie
N	4 speed	Muncie
N-P	4 speed	Muncie
O-D	3 speed OD	Saginaw
P	4 speed	Muncie (El Camino)
R	4 speed	Saginaw
S	3 speed	Saginaw
T	Powerglide	Toledo
W	3/4 speed	Warner Gear

Axle Identification

Code	Ratio
El Camino	
CQ, CS, KE, KI	2.56:1
CH, CL, CM, CZ	
KD, KQ, KR, KW	
KC	2.73:1
CD, CJ, CK, CX	
CY, KS, KT, KX	3.07:1
CA, CE,	3.08:1
CF, CN, CO, CW	
KU, KV, KY, KZ	3.31:1
CB, CG,	3.36:1
KA, KB, KH, KG	
KJ	3.55:1
CR, CV	3.70:1
CC, CI, CR, CT	
CU,	3.73:1
KK, KL	4.10:1
KM, KN	4.56:1
KO, KP	4.88:1

C-10 Series

HB	3.07:1
HA, HC, MD	3.73:1
HD, JQ,	4.11:1

C-20 Series

HW, JU	4.11:1
HU, HV, HX, HY, HZ, JS, JT	4.57:1

C-30 Series

LW, JN, MC, MW MY, MP	4.57:1
LU, IC, ID, IE, IF LU, LV, MO, MS, MV, MQ, MT, MU	5.14:1
MX	6.17:1

K-10 Series

HQ	3.73:1

K-20 Series

HZ, ND	4.57:1

Exterior Color Codes

C-10, C-20, C-30 Series

Black	500
Light Green	503
Dark Green	505
Medium Blue	506
Light Blue	507
Dark Blue	508
Dark Aqua	511
Red	514
Vermillion	515
Orange	516
Dark Yellow	519
White	521
Silver	523
Ivory	526

El Camino

Tuxedo Black	AA
Ermine White	CC
Nantucket Blue	DD
Deepwater Blue	EE
Marina Blue	FF
Granada Gold	GG
Mountain Green	HH
Emerald Turquoise	KK
Tahoe Turquoise	LL
Royal Plum	MM
Madeira Maroon	NN
Bolero Red	RR
Sierra Fawn	SS
Capri Cream	TT
Butternut Yellow	YY

Regular Production Options

El Camino

13480	El Camino 8cyl	$2,613
13680	Custom El Camino	2,694

C-10 Series

CE10702	Chassis and cowl	
CE10703	Chassis and cab	2,175
CE10704	Stepside pickup	2,301
CE10734	Fleetside pickup	2,339
CE10904	Stepside pickup	2,339
CE10905	Panel delivery	2,757
CE10906	Suburban w/rear doors	2,988
CE10934	Fleetside pickup	2,377

C-20 Series

CE20903	Chassis and cab	2,352
CE20904	Stepside pickup	2,352
CE20905	Panel delivery	2,897
CE20906	Suburban w/rear doors	3,124
CE20909	Stake bed	2,568
CE20934	Fleetside pickup	2,516

C-30 Series

CS31003	Chassis & cab	2,561
CS31004	Stepside pickup	2,695
CS31009	Stake bed	2,875
CS31034	Stepside pickup	2,755

El Camino

Option Number

AS1	Std shoulder harness
A01	Soft ray tinted glass
A02	Soft ray tinted glass (windshield only)
A31	Power windows
A41	Four way electric power seat
A49	Custom deluxe seat belts w/retractors
A51	Strato-Bucket seats
A81	Strato-Ease headrests
A82	Strato-Ease headrests
A85	Deluxe shoulder harness
B37	Floor mats
B55	Deluxe front seat cushion
B90	Side window moldings
B93	Door edge guards
C08	Vinyl roof cover
C48	Heater & defroster deletion
C50	Rear window defroster
C60	Four season A/C
D33	Remote control outside LH mirror
D55	Front compartment console
F40	Special front & rear suspension
G66	Superlift rear shock absorbers
G75	3.70:1 rear axle ratio
G76	3.36:1 rear axle ratio
G80	Positraction rear axle
G92	3.08:1 rear axle ratio
G94	3.31:1 rear axle ratio
G96	3.55:1 rear axle ratio
G97	2.73:1 rear axle ratio
J50	Power brakes
J52	Front disc brakes
J65	Special brakes w/metallic facings
K02	Temperature controlled radiator fan
K19	AIR equipment
K24	Closed engine positive engine ventilation
K30	Speed & cruise control
K76	61 amp Delcotron generator
K79	42 amp Delcotron generator
L22	155hp Turb-Thrift six cylinder engine
L30	275hp Turbo-Fire 327ci engine
L78	350hp Turbo-Jet 396ci engine
L79	325hp Turbo-Fire 327ci engine
M01	Heavy duty clutch
M10	Overdrive transmission
M13	Special fully synchronized 3 speed transmission
M20	Four speed wide range transmission
M21	Four speed close ratio transmission
M22	Four speed transmission heavy duty
M35	Powerglide transmission

M40	Turbo-Hydramatic transmission
N10	Dual exhaust
N30	Deluxe steering wheel
N33	Comfortilt steering wheel
N34	Sports styled steering wheel
N40	Power steering
N96	Simulated magnesium wheel trim cover
PO1	4 brightmetal wheel covers
P02	Simulated wire wheel covers
T60	Heavy duty battery
U03	Tri-Volume horn
U15	Speed warning indicator
U14	Special instrumentation
U16	Tachometer
U26	Underhood lamp
U27	Glove compartment lamp
U28	Ashtray lamp
U29	Instrument panel courtesy lights
U35	Electric clock
U57	Stereo tape system
U63	Radio—push button control
U69	Radio AM/FM push button control
V01	HD radiator
V31	Front bumper guard
V32	Rear bumper guard
Z29	Special bodyside accent stripes

C-10, C-20, C-30 Series

Option Number

AO9	Laminated glass
A10	Full view rear window
A11	Soft ray tinted glass
A12	Rear window glass
A13	Side door glass
A37	Seat belts, custom deluxe
A54	Full width front seat
A55	Bostrom seat
A57	Auxiliary seat
A59	Supplementary seat
A62	Front seat belt (delete)
A97	Right door lock
B78	Dispatch box door equipment
B98	Custom side molding
C14	Electric windshield wipers & washer
C40	Direct-air heater & defroster
C41	Thrift-air heater & defroster
C60	Air conditioning
D29	West coast mirror—Jr
D30	West coast mirror—Sr
D32	Rear view mirror
E56	Platform and stake rack

E57	Platform body
E80	Pickup box mounting brackets
E82	Level box pickup floor
E85	Body side door equipment
FO3	Heavy duty frame
F49	Heavy duty four wheel drive front axle
F51	Front and rear HD shock absorbers
F60	Heavy duty front springs
F76	Front wheel locking hubs
F81	Front springs, HD
G50	Heavy duty rear springs
G52	7,500lb rear springs
G55	8,750lb rear springs
G60	Auxiliary rear springs
G80	Limited slip rear axle
G81	Positraction differential
G82	No-Spin rear axle
J70	Vacuum brakes
J71	Full air brakes
J72	Air-hydraulic brakes
J73	Heavy duty vacuum brakes
J80	Vacuum brake reserve tank
J81	Vacuum gauge
J91	Trailer air brake equipment
K12	Oil filter
K19	AIR equipment
K23	Positive crankcase ventilation
K24	Closed engine positive engine ventilation type B
K28	Fuel filter equipment
K31	Throttle control
K37	Engine governor
K47	Air cleaner equipment
K48	Oil bath air cleaner
K56	Air compressor equipment
K67	Heavy duty starter motor
K71	35 amp low cut-in DC generator
K77	55 amp Delcotron generator
K79	42 amp Delcotron generator
K81	62 amp Delcotron generator
L05	130 amp Delcotron generator
M01	Heavy duty clutch
M55	Powerglide transmission
NO2	Fuel tank—30 gallon
N12	Exhaust system—single stack
N13	Exhaust system—dual stack
N34	Sports-styled walnut grained steering wheel w/plastic ring
N40	Power steering
PO1	Wheel trim cover
P10	Spare wheel carrier—under frame mounting
P13	Spare wheel carrier—side mounted
T60	Heavy duty battery
U16	Tachometer

129

Facts

The 1967 truck line was restyled, exhibiting cleaner, more integrated, and modern styling. The cab and hood were lower in profile, which lowered the truck's center of gravity and contributed to better handling. Collectively, the 1967—72 trucks are known as the Custom Sports Trucks (CSTs) because of the CST option.

Mechanically, the chassis and drivetrain were basically carry-overs. The 250ci was the base engine, rated at 140hp. Optional were the larger 292ci six at 170hp and the 283ci V-8 rated at 175hp. The largest engine option was the 327ci V-8 rated at 220hp. The sixes came with a one-barrel carburetor, the 283ci V-8 got a two-barrel, and the 327ci V-8 came with a four-barrel.

The standard transmission was the three-speed manual; a three-speed overdrive manual, a four-speed manual, and the Powerglide and Turbo Hydra-matic automatics were optional. The three-speed overdrive was available only on the 1/2-ton trucks.

As before, the suspension used coil springs at all four corners on the 1/2- and 3/4-ton models and leaf springs at the rear on the 1-ton trucks. The front suspension was independent. Four-wheel-drive-equipped trucks used leaf springs instead of coils and were all equipped with a manual transmission.

Two basic trim levels were available—standard and Custom—along with a Custom variation that included, for the first time, bucket seats. The Custom trim included a chrome front bumper, a chrome grille, chrome window and door moldings, and a higher level of interior accoutrements.

The CST option consisted of a chrome front bumper, a chrome grille, brightmetal window moldings, and a full-length brightmetal molding that ran along the truck's side, following the curve of both wheel arches. In the interior, the CST option included additional bright trim on the dash and doors, bright pedal trim, a full-width vinyl seat, carpeting, a door-operated dome light, and a cigar lighter. On the exterior, a CST emblem appeared on both doors. The CST option was available only on trucks with the Fleetside-type box. Only 12,588 trucks were so equipped.

The Suburban was now equipped with an additional passenger's-side door.

The El Camino was mildly facelifted, using different grille and rear end treatments. Although no El Camino SS was offered, the L35 325hp 396ci V-8 could be had as an option in the El Camino, as could the L34 396ci rated at 350hp.

Front disc brakes were optional for the first time on the El Camino but required the use of the optional Rally wheels.

The dual brake master cylinder was made standard equipment on the 1967 El Camino. The three-speed Turbo Hydra-matic automatic transmission was available for the first time this year.

Chapter 24

1968 Trucks

Production

13380 El Camino 6cyl	2,757
13480 El Camino 8cyl	6,082
13580 El Camino Custom 6cyl	1,072
13680 El Camino Custom 8cyl	26,690
13880 El Camino SS 396	5,190
Total El Camino	41,791

C-10 Series   K Series

CE10703 Chassis and cab	2,735	43
CE10704 Stepside pickup	46,322	1,706
CE10734 Fleetside pickup	46,483	1,449
CE10904 Stepside pickup	18,632	552
CE10905 Panel delivery	4,801	59
CE10906 Suburban w/rear doors	11,004	1,143
CE10934 Fleetside pickup	204,236	3,625

C-20 Series

CE20903 Chassis and cab	6,636	498
CE20904 Stepside pickup	7,666	1,047
CE20905 Panel delivery	1,572	68
CE20906 Suburban w/rear doors	1,573	299
CE20909 Stake bed	1,103	—
CE20934 Fleetside pickup	60,646	4,704

C-30 Series

CE31002 Chassis & cowl	238
CS31003 Chassis & cab	11,948
CS31004 Stepside pickup	2,836
CS31009 Stake bed	3,272
CE31034 Stepside pickup	213
CE31403 Chassis & cab	5,639
1968 Model Year Production: 1/2, 3/4, 1 ton	521,486

Serial numbers

Description

CE107048A100001

C—Chassis type, C—conventional, K—four-wheel drive

E—Engine: E—V-8, S—V-6

1—GVW range: 1—3600-5600lb, 2—5500-8100lb, 3—6700-10,000lb

07—Cab to axle dimension: 07—42-47", 09—54-59", 10—60-65", 14—84-89"

04—Body type, 02—Cowl, 03—Cab, 04—Stepside pickup, 05—Panel, 06—
    Suburban (panel rear doors), 09—Platform stake, 12—Windshield cowl,

13—Cab (full air brakes), 16—Suburban (tail & liftgate), 34—Fleetside pickup

8—Last digit of model year, 8—1968

A—Assembly Plant Code: A—Atlanta GA, B—Baltimore MD, F—Flint MI, J—Janesville WI, K—Kansas City MO, S—St. Louis MO, T—Tarrytown NY, Z—Fremont CA, 1—Oshawa Canada

100001—Consecutive Sequence Number

Location: On plate attached to left door hinge post. On cowl models, plate is attached to left cowl inner panel.

El Camino Serial Number

136808A100001

1—Chevrolet Division

36—Series, 33—El Camino 6cyl, 34—El Camino 8cyl, 35—Custom El Camino 6cyl, 36—Custom El Camino 8cyl, 38—El Camino SS 396

80—Body style, 2 door sedan pickup

8—Last digit of model year, 8—1968

A—Assembly Plant Code: A—Atlanta GA, B—Baltimore MD, G—Framingham MA, K—Kansas City MO, Z—Fremont CA

100001—Consecutive Sequence Number

Location: On plate attached to driver's side of dash, visible through the windshield.

Model & Wheelbase

Model Number and Description	Wheelbase (in)
El Camino	
13380 El Camino 6cyl	115
13480 El Camino 8cyl	115
13580 El Camino Custom 6cyl	115
13680 El Camino Custom 8cyl	115
13880 El Camino SS 396	115
C-10 Series	
CE10702 Chassis and cowl	115
CE10703 Chassis and cab	115
CE10704 Stepside pickup	115
CE10734 Fleetside pickup	115
CE10903 Chassis & cab	127
CE10904 Stepside pickup	127
CE10905 Panel delivery	127
CE10906 Suburban w/rear doors	127
CE10916 Suburban w/tailgate	127
CE10934 Fleetside pickup	127
C-20 Series	
CE20902 Chassis and cowl	127
CE20903 Chassis and cab	127
CE20904 Stepside pickup	127
CE20905 Panel delivery	127
CE20906 Suburban w/rear doors	127
CE20909 Stake bed	127

CE20912	Windshield Cowl	127
CE20916	Suburban w/tailgate	127
CE20934	Longhorn pickup	133

C-30 Series

CE31002	Chassis & cowl	133
CE31003	Chassis & cab	133
CE31004	Stepside pickup	133
CE31009	Stake bed	133
CE31034	Longhorn pickup	133
CE31403	Chassis & cab	157

K-10 Series

KE10703	Chassis and cab	115
KE10704	Stepside pickup	115
KE10734	Fleetside pickup	115
KE10903	Chassis & cab	127
KE10904	Stepside pickup	127
KE10905	Panel delivery	115
KE10906	Suburban w/rear doors	127
KE10916	Suburban w/tailgate	127
KE10934	Fleetside pickup	127

K-20 Series

KE20903	Chassis and cab	127
KE20904	Stepside pickup	127
KE20905	Panel delivery	127
KE20906	Suburban w/rear doors	127
KE20916	Suburban w/tailgate	127
KE20934	Fleetside pickup	127

Engine & Transmission Suffix Codes

El Camino

BA, BC—230ci I-6 1bbl 140hp—manual trans
BB, BD—230ci I-6 1bbl 140hp—manual trans
BF—230ci I-6 1bbl 140hp—Powerglide
BH—230ci I-6 1bbl 140hp—Powerglide W/C.A.C.
CM, CN—250ci I-6 1bbl 155hp—Manual trans
CQ, CR—250ci I-6 1bbl 155hp—Powerglide
DA, DB—307ci V-8 2bbl 200hp—manual trans
DE, DN—307ci V-8 2bbl 200hp—Powerglide
EA, EO—327ci V-8 2bbl 250hp—manual trans
EE—327ci V-8 2bbl 250hp—Powerglide
EC—327ci V-8 4bbl 275hp—Powerglide
EH, EJ—327ci V-8 4bbl 275hp—manual trans
EI—327ci V-8 4bbl 275hp—Powerglide
EP, ES—327ci V-8 4bbl 325hp—Manual trans
ED—396ci V-8 4bbl 325hp—Three or four-speed manual
EK—396ci V-8 4bbl 325hp—Powerglide automatic
ET—396ci V-8 4bbl 325hp—Turbo-Hydramatic automatic
EF—396ci V-8 4bbl 350hp—Three or four-speed manual
EL—396ci V-8 4bbl 350hp—Powerglide automatic
EU—396ci V-8 4bbl 350hp—Turbo-Hydramatic automatic
EG—396ci V-8 4bbl 375hp—Four-speed manual

C-10, C-20 Series

PC—250ci I-6 1bbl 155hp—manual trans*
PF—250ci I-6 1bbl 155hp—four-speed manual, HD
PW—250ci I-6 1bbl 155hp—three-speed manual trans*
PX—250ci I-6 1bbl 155hp—four-speed manual
TA—250ci I-6 1bbl 155hp—manual trans
TC—250ci I-6 1bbl 155hp—manual trans
TD—250ci I-6 1bbl 155hp—manual
TE—250ci I-6 1bbl 155hp—automatic
TF—250ci I-6 1bbl 155hp—four-speed manual
TT—250ci I-6 1bbl 155hp—automatic
UH—292ci I-6 1bbl 170hp—manual trans
UM—292ci I-6 1bbl 170hp—manual trans*
UN—292ci I-6 1bbl 170hp—manual trans*
UO—292ci I-6 1bbl 170hp—Powerglide*
UR—292ci I-6 1bbl 170hp—manual trans
UT—292ci I-6 1bbl 170hp—Powerglide
U3—292ci I-6 1bbl 170hp—manual trans*
WA—307ci V-8 2bbl 200hp—manual trans
WB—307ci V-8 2bbl 200hp—manual trans*
WC—307ci V-8 2bbl 200hp—three-speed manual
WE—307ci V-8 2bbl 200hp—Powerglide
WF—307ci V-8 2bbl 200hp—manual trans*
WG—307ci V-8 2bbl 200hp—manual trans*
WH—307ci V-8 2bbl 200hp—manual trans
WO—307ci V-8 2bbl 200hp—three-speed manual
WR—307ci V-8 2bbl 200hp—Powerglide*
ZO—307ci V-8 2bbl 200hp—manual trans
ZS—307ci V-8 2bbl 200hp—manual trans*
YB—327ci V-8 4bbl 220hp—Powerglide
YC—327ci V-8 4bbl 220hp—manual trans
YD—327ci V-8 4bbl 220hp—Powerglide
YK—327ci V-8 4bbl 220hp—three-speed manual
YH—327ci V-8 4bbl 220hp—three-speed manual*
YR—327ci V-8 4bbl 220hp—Powerglide*
YS—327ci V-8 4bbl 220hp—manual trans*
YV—327ci V-8 4bbl 220hp—manual trans
XE—396ci V-8 4bbl 310hp—manual trans/AIR
XF—396ci V-8 4bbl 310hp—manual trans
XG—396ci V-8 4bbl 310hp—three-speed manual
XH—396ci V-8 4bbl 310hp—three-speed manual*
XK—396ci V-8 4bbl 275hp—manual trans
XL—396ci V-8 4bbl 275hp—three-speed manual
XM—396ci V-8 4bbl 275hp—three-speed manual*
XN—396ci V-8 4bbl 275hp—manual trans*
*C-20

C-30 Series

PW—250ci I-6 1bbl 155hp—three-speed manual
PX—250ci I-6 1bbl 155hp—four-speed manual
TF—250ci I-6 1bbl 155hp—four-speed manual/HD
TL—250ci I-6 1bbl 155hp—manual trans
UM—292ci I-6 1bbl 170hp—manual trans
UN—292ci I-6 1bbl 170hp—manual trans
UO—292ci I-6 1bbl 170hp—three-speed manual
UQ—292ci I-6 1bbl 170hp—manual trans
WB—307ci V-8 2bbl 200hp—manual trans

WG—307ci V-8 2bbl 200hp—manual trans
WO—307ci V-8 2bbl 200hp—three-speed manual
WR—307ci V-8 2bbl 200hp—three-speed manual
YH—327ci V-8 4bbl 220hp—three-speed manual
YS—327ci V-8 4bbl 220hp—manual trans
XF—396ci V-8 4bbl 310hp—manual trans
XN—396ci V-8 4bbl 275hp—manual trans

K-10 Series
PF—250ci I-6 1bbl 155hp—manual trans
TA—250ci I-6 1bbl 155hp—four-speed manual
TT—250ci I-6 1bbl 155hp—Powerglide
UH—292ci I-6 1bbl 170hp—manual trans
UR—292ci I-6 1bbl 170hp—manual trans
WI—307ci V-8 2bbl 200hp—manual trans
ZX—307ci V-8 2bbl 200hp—three-speed manual
ZP—307ci V-8 2bbl 200hp—manual trans
YX—327ci V-8 4bbl 220hp—manual trans
YT—327ci V-8 4bbl 220hp—three-speed manual

K-20 Series
PF—250ci I-6 1bbl 155hp—four-speed manual
PV—250ci I-6 1bbl 155hp—manual trans
TT—250ci I-6 1bbl 155hp—Powerglide
TA—250ci I-6 1bbl 155hp—manual trans
TC—250ci I-6 1bbl 155hp—manual trans
TD—250ci I-6 1bbl 155hp—manual trans
UH—292ci I-6 1bbl 170hp—manual trans
U3—292ci I-6 1bbl 170hp—manual trans
UM—292ci I-6 1bbl 170hp—manual trans
UN—292ci I-6 1bbl 170hp—manual trans
UR—292ci I-6 1bbl 170hp—manual trans
WI—307ci V-8 2bbl 200hp—manual trans
WH—307ci V-8 2bbl 200hp—manual trans
ZY—307ci V-8 2bbl 200hp—three-speed manual
ZX—307ci V-8 2bbl 200hp—three-speed manual
ZP—307ci V-8 2bbl 200hp—manual trans
ZQ—307ci V-8 2bbl 200hp—manual trans
YL—327ci V-8 4bbl 220hp—manual trans
YT—327ci V-8 4bbl 220hp—three-speed manual
YU—327ci V-8 4bbl 220hp—three-speed manual

Transmission Codes

Code	Type	Plant
C	Powerglide	Cleveland
CA	Hydra-Matic	—
E	Powerglide	McKinnon
K	3 speed	McKinnon
M	3 speed O/D	Muncie
N	4 speed	Muncie
N-P	4 speed	Muncie
O	3 speed O/D	Saginaw
P	4 speed	Muncie
R	4 speed	Saginaw
S	3 speed	Saginaw
T	Powerglide	Toledo
W	3/4 speed	Warner Gear

Axle Identification

Code / Ratio

El Camino

Code	Ratio
CQ, CS, KE, KI	2.56:1
KR	2.72:1
CH, CL, CP, CZ	
FH, FI, KC, KD	
KW	2.73:1
CD, CJ, CK, CX	
CY, KS, KT, KX	3.07:1
CA, CE, FJ, FK	3.08:1
CF, CN, CO, CW	
KU, KV, KY, KZ	3.31:1
CB, CG, FL, FM	3.36:1
FN, FO, KA, KB	
KF, KH, K2, K3	3.55:1
CR, CV, FP, FQ	
CC, CI, CR, CT	
CU, KR, K5	3.73:1
KK, KL, K6	4.10:1
KM, KN, K7	4.56:1
KO, KP, K8	4.88:1

C-10 Series

Code	Ratio
HB, LL, RA, RT,	
RU, RV, RW	
TR, TT, WH	3.07:1
HC, HI, TU	3.36:1
HA, HQ	3.73:1
HQ, JQ, TS, TV	4.11:1

C-20 Series

Code	Ratio
NA, NB, OT, OW	
OY, WI	4.10:1
HW, JV	4.11:1
OX, OZ	4.56:1
HU, HV, HX, HY	
JS, JT, OU	4.57:1

C-30 Series

Code	Ratio
RN	4.10:1
IE, LJ, LX, LZ	
NZ	4.11:1
IG, LW, MC, NN	
VL, VM, VN, VO	
LU, IC, ID, IE, IF	
ID, IF, LV, LY	
TX, TY, VF, VS	5.14:1
MX	6.17:1

K-10 Series

Code	Ratio
NX	3.07:1
HQ	3.73:1

K-20 Series

Code	Ratio
NC, ND, NE	4.57:1

Exterior Color Codes

C-10, C-20, C-30 Series

Color	Code
Black	500
Light Green	503
Dark Green	505
Medium Blue	506
Light Blue	507
Dark Blue	508
Red	514
Vermillion	515
Orange	516
Dark Yellow	519
Light Yellow	520
White	521
Silver	523
Saddle Fawn	525
Ivory	526

El Camino

Color	Code
Tuxedo Black	AA
Ermine White	CC
Grotto Blue	DD
Fathom Blue	EE
Island Teal	FF
Ash Gold	GG
Grecian Green	HH
Tripoli Turquoise	KK
Teal Blue	LL
Cordovan Maroon	NN
Seafrost Green	PP
Matador Red	RR
Palomino Ivory	TT
Sequoia Green	VV
Butternut Yellow	YY

Regular Production Options

El Camino

13380	El Camino 6cyl	$2,482
13580	Custom El Camino 6cyl	2,563
13480	El Camino	2,590
13680	Custom El Camino	2,671
13880	SS 396 El Camino	2,926

C-10 Series

CE10703	Chassis and cab	2,320
CE10704	Stepside pickup	2,430
CE10734	Fleetside pickup	2,468
CE10904	Stepside pickup	2,468
CE10905	Panel delivery	2,839
CE10906	Suburban w/rear doors	3,081
CE10934	Fleetside pickup	2,647

C-20 Series

CE20903	Chassis and cab	2,499
CE20904	Stepside pickup	2,610
CE20905	Panel delivery	2,981
CE20906	Suburban w/rear doors	3,264
CE20909	Stake bed	2,702
CE20934	Fleetside pickup	2,647

C-30 Series

CS31003	Chassis & cab	2,561
CS31004	Stepside pickup	2,695
CS31034	Stepside pickup	2,755

El Camino

Option Number

AS1	Front seatbelts & shoulder belts
A01	Soft ray tinted glass
A02	Soft ray tinted glass (windshield only)
A31	Power windows
A39	Custom deluxe seatbelts
A51	Strato-Bucket seats
A81	Headrests w/Strato-buckets
A82	Headrests w/full width front seat
A85	Deluxe deluxe shoulder belts
B37	Color keyed floor mats
B55	Extra thick foam seat cushion
B90	Side window moldings
B93	Door edge guards
C08	Vinyl roof cover
C24	Hide-a-Way windshield wipers
C50	Rear window defroster
C60	Four season A/C
D33	Remote control outside LH mirror
D55	Console
D96	Accent stripping
F40	Special front & rear suspension
G66	Superlift rear shock absorbers
G75	3.70:1 rear axle ratio
G76	3.36:1 rear axle ratio
G80	Positraction rear axle
G92	3.08:1 rear axle ratio
G94	3.31:1 rear axle ratio
G96	3.55:1 rear axle ratio
G97	2.73:1 rear axle ratio
HO1	3.07:1 rear axle ratio
HO5	3.73:1 rear axle ratio
J50	Power brakes
J52	Front disc brakes
KD5	HD engine ventilation
K02	Temperature controlled radiator fan
K30	Cruise-Master speed control
K76	61 amp Delcotron generator
K79	42 amp Delcotron generator
L22	155hp Turbo-Thrift six cylinder engine
L30	275hp Turbo-Fire 327ci engine
L34	350hp Turbo-Jet 396ci engine
L73	250hp Turbo-Fire 327ci engine
L78	375hp Turbo-Jet 396ci engine
L79	325hp Turbo-Fire 327ci engine
M01	Heavy duty clutch
M10	Overdrive transmission
M13	Special fully synchronized 3 speed transmission
M20	Four speed wide range transmission
M21	Four speed close ratio transmission
M22	Four speed transmission heavy duty
M35	Powerglide transmission
M40	Turbo-Hydramatic transmission
N10	Dual exhaust
N30	Deluxe steering wheel
N33	Comfortilt steering wheel
N34	Sports styled steering wheel
N40	Power steering
E85	Body side door equipment
FO3	Heavy duty frame
F49	Heavy duty four wheel drive front axle
F51	Front and rear HD shock absorbers
F57	9000lb front axle
F58	11,000lb front axle
F59	Front stabilizer bar
F60	Heavy duty front springs
F76	Front wheel locking hubs
F81	Front springs, HD
G50	Heavy duty rear springs
G52	7,500lb rear springs
G55	8,750lb rear springs
G56	10,400lb rear springs
G60	Auxiliary rear springs
G80	Limited slip rear axle
G81	Positraction differential
G82	No-Spin rear axle
J70	Vacuum brakes
J71	Full air brakes
J72	Air-hydraulic brakes
J73	Heavy duty vacuum brakes
J80	Vacuum brake reserve tank
J81	Vacuum gauge
A09	Laminated glass
A10	Full view rear window
A11	Soft ray tinted glass
A12	Rear window glass
A13	Side door glass
A37	Seat belts, custom deluxe

A54	Full width front seat	K37	Engine governor
A55	Bostrom seat	K47	Air cleaner equipment
A57	Auxiliary seat	K48	Oil bath air cleaner
A59	Supplementary seat	K56	Air compressor equipment
A62	Front seat belt (delete)	K67	Heavy duty starter motor
A97	Right door lock	K71	35 amp low cut-in DC generator
B78	Dispatch box door equipment	K77	55 amp Delcotron generator
B98	Custom side molding	K79	42 amp Delcotron generator
C14	Electric windshield wipers & washer	K81	62 amp Delcotron generator
C40	Direct-air heater & defroster	L05	130 amp Delcotron generator
C41	Thrift-air heater & defroster	M01	Heavy duty clutch
C60	Air conditioning	M16	Warner T89B three speed transmission
D29	West coast mirror—Jr	M49	Powermatic transmission
D30	West coast mirror—Sr	M55	Powerglide transmission
D32	Rear view mirror	NO2	Fuel tank—30 gallon
E56	Platform and stake rack	N12	Exhaust system—single stack
E57	Platform body	N13	Exhaust system—dual stack
E80	Pickup box mounting brackets	N34	Sports-styled walnut grained steering wheel w/plastic ring
E82	Level box pickup floor	N40	Power steering
N95	Simulated wire wheel covers	PO1	Wheel trim cover
N96	Simulated magnesium wheel trim cover	P10	Spare wheel carrier—under frame mounting
PA2	Mag-spoke wheel covers	P13	Spare wheel career—side mounted
T60	Heavy duty battery	T60	Heavy duty battery
U03	Tri-Volume horn	U16	Tachometer
U14	Special instrumentation	U60	Radio—manual
U15	Speed warning indicator	V01	HD radiator
U14	Special instrumentation	V04	Radiator shutters
U26	Underhood lamp	V05	HD cooling
U27	Glove compartment lamp	V35	Wraparound front bumper
U28	Ashtray lamp	V37	Custom chrome option
U29	Instrument panel courtesy lights	V38	Rear bumper—painted
U35	Electric clock	V62	Hydraulic jack
U46	Light monitoring system	V75	Hazard and marker lights
U57	Stereo tape system	V76	Front tow hooks
U63	Radio—push button control	Z02	Push button control radio & rear speaker
U69	Radio AM/FM push button control	Z12	Speedometer, driven gear & fitting
U79	AM/FM stereo radio	Z50	Frame reinforcements
V01	HD radiator	Z52	Full depth foam seat
V31	Front bumper guards	Z53	Gauges
V32	Rear bumper guards	Z54	Maximum economy equipment
ZJ7	Rally wheels	Z56	15,000lb GVW plate

C10, C20, C30 Series

AO9	Laminated glass	Z57	23,000lb GVW plate
J91	Trailer air brake equipment	Z59	21,000lb GVW plate
K12	Oil filter	Z60	Custom equipment
K19	AIR equipment	Z62	Custom comfort and appearance option
K23	Positive crankcase ventilation	Z70	7,800lb GVW plate
K24	Closed engine positive engine ventilation type B	Z72	Vacuum equipment
K28	Fuel filter equipment	Z73	5,000lb GVW plate
K31	Throttle control	Z81	Camper special equipment
		Z84	Custom Sport Truck package

Facts

Visually, the 1968 trucks got federally mandated side marker lights on the front and rear fenders. They also received a silver anodized front grille panel and headlight receptacle area when an optional exterior trim package was ordered.

The standard dog-dish hubcap was painted white with a red bow-tie surrounded by a red ring. In 1967, the ring was black.

The CST option was still available, and 16,755 trucks were so equipped.

A Custom comfort and appearance option included brightmetal window moldings, color-keyed vinyl floor mats, foam seats with color-keyed trim, armrests, chrome knobs, and cowl insulation. Trucks so equipped were also awarded Custom front fender emblems.

The 283ci V-8 was retired and replaced by a 307ci version rated at 200hp. Also new on the options list were two versions of the 396ci big-block: a two-barrel rated at 275hp and a four-barrel rated at 310hp.

The Chevelle line was restyled, and so too the El Camino. For the first time, the factory offered an El Camino SS model, powered by the 325hp 396ci V-8. Optional were 350hp or 375hp versions.

SS features were a blacked-out grille with an SS 396 emblem, a blacked-out taillight panel with an SS 396 emblem, a black lower body area on light-colored cars, a dome hood, and lower rear fender moldings. The numberals 396 were incorporated on the front fender—mounted signal lamps. Early cars included SS letters with the 396 numerals.

Full-size SS wheel covers were optional. In the interior, an SS emblem was located on the dash panel above the glovebox door.

Beginning with 1968, the VIN plate was located on the dash and visible through the windshield on all El Caminos.

Hidden windshield wipers were standard on the SS 396.

El Camino 396s got improved finned brake drums in the front. Optional front disc brakes were available with the Rally wheels.

From this year on, all engines used a spin-on oil filter rather than the previous canister setup.

Chapter 25

1969 Trucks

Production

13380 El Camino 6cyl	2,193
13480 El Camino 8cyl	6,407
13580 El Camino Custom 6cyl	785
13680 El Camino Custom 8cyl	39,000
Total El Camino	48,385

C-10 Series / K Series

		K Series
CE10703 Chassis and cab	2,343	58
CE10704 Stepside pickup	49,147	1,698
CE10734 Fleetside pickup	54,211	1,849
CE10904 Stepside pickup	18,170	521
CE10905 Panel delivery	5,492	—
CE10906 Suburban w/rear doors	7,263	697
CE10934 Fleetside pickup	268,233	4,937

C-20 Series

CE20903 Chassis and cab	8,440	556
CE20904 Stepside pickup	8,090	1,071
CE20905 Panel delivery	1,779	—
CE20906 Suburban w/rear doors	1,649	289
CE20916 Suburban w/rear gate	1,087	256
CE20934 Fleetside pickup	74,894	6,124
CE21034 Fleetside pickup	8,797	—

C-30 Series

CE31002 Chassis & cowl	236
CS31003 Chassis & cab	16,828
CS31004 Stepside pickup	2,457
CE31034 Stepside pickup	8,797
CE31403 Chassis & cab	8,290

K-5 Blazer

KS10514	4,935
1968 Model Year Production: ½, ¾, 1 ton	643,622

Serial numbers

Description

CE107049A100001
C—Chassis type, C—conventional, K—four-wheel drive
E—Engine: E—V-8, S—V-6
1—GVW range: 1—3600-5600lb, 2—5500-8100lb, 3—6700-10,000lb
07—Cab to axle dimension: 07—42-47", 09—54-59", 10—60-65", 14—84-89"
04—Body type, 02—Cowl, 03—Cab, 04—Stepside pickup, 05—Panel,

06—Suburban (panel rear doors), 09—Platform stake, 12—Windshield cowl, 14—Utility Blazer, 16—Suburban (tail & liftgate), 34—Fleetside pickup

9—Last digit of model year, 9—1969

A—Assembly Plant Code: A—Atlanta GA, B—Baltimore MD, F—Flint MI, J—Janesville WI, K—Kansas City MO, S—St. Louis MO, T—Tarrytown NY, Z—Fremont CA, 1—Oshawa Canada

100001—Consecutive Sequence Number

Location: On plate attached to left door hinge post. On cowl models, plate is attached to left cowl inner panel.

El Camino Serial Number

136809A100001

1—Chevrolet Division

36—Series, 33—El Camino 6cyl, 34—El Camino 8cyl, 35—Custom El Camino 6cyl, 36—Custom El Camino 8cyl

80—Body style, 2 door sedan pickup

9—Last digit of model year, 9—1969

A—Assembly Plant Code: A—Atlanta GA, B—Baltimore MD, G—Framingham MA, K—Kansas City MO, Z—Fremont CA

300001—Consecutive Sequence Number

Location: On plate attached to driver's side of dash, visible through the windshield.

Model & Wheelbase

Model Number and Description	Wheelbase (in)
El Camino	
13380 El Camino 6cyl	115
13480 El Camino 8cyl	115
13580 El Camino Custom 6cyl	115
13680 El Camino Custom 8cyl	115
C-10 Series	
CE10702 Chassis and cowl	115
CE10703 Chassis and cab	115
CE10704 Stepside pickup	115
CE10734 Fleetside pickup	115
CE10903 Chassis & cab	127
CE10904 Stepside pickup	127
CE10905 Panel delivery	127
CE10906 Suburban w/rear doors	127
CE10916 Suburban w/tailgate	127
CE10934 Fleetside pickup	127
C-20 Series	
CE20902 Chassis and cowl	127
CE20903 Chassis and cab	127
CE20904 Stepside pickup	127
CE20905 Panel delivery	127
CE20906 Suburban w/rear doors	127
CE20909 Stake bed	127
CE20912 Windshield Cowl	127

| CE20916 Suburban w/tailgate | 127 |
| CE20934 Longhorn pickup | 133 |

C-30 Series
CE31002 Chassis & cowl	133
CE31003 Chassis & cab	133
CE31004 Stepside pickup	133
CE31009 Stake bed	133
CE31034 Longhorn pickup	133
CE31403 Chassis & cab	157

K-5 Series
| KE10514 Blazer | 104 |

K-10 Series
KE10703 Chassis and cab	115
KE10704 Stepside pickup	115
KE10734 Fleetside pickup	115
KE10903 Chassis & cab	127
KE10904 Stepside pickup	127
KE10905 Panel delivery	115
KE10906 Suburban w/rear doors	127
KE10916 Suburban w/tailgate	127
KE10934 Fleetside pickup	127

K-20 Series
KE20903 Chassis and cab	127
KE20904 Stepside pickup	127
KE20905 Panel delivery	127
KE20906 Suburban w/rear doors	127
KE20916 Suburban w/tailgate	127
KE20934 Fleetside pickup	127

Engine & Transmission Suffix Codes

El Camino
AM, AD—230ci I-6 1bbl 140hp—manual trans
AQ—230ci I-6 1bbl 140hp—manual trans
AN, AP—230ci I-6 1bbl 140hp—Powerglide
AR—230ci I-6 1bbl 140hp—Powerglide
BB, BD—250ci I-6 1bbl 155hp—Manual trans
BF—250ci I-6 1bbl 155hp—manual trans
BC, BE—250ci I-6 1bbl 155hp—Powreglide
BH—250ci I-6 1bbl 155hp—Powerglide
DA, DE—307ci V-8 2bbl 200hp—manual trans
DC, DD—307ci V-8 2bbl 200hp—Powerglide
HC, HS—350ci V-8 4bbl 255hp—manual trans
HR—350ci V-8 4bbl 255hp—Powerglide
EE—350ci V-8 4bbl 250hp—Powerglide
JA—396ci V-8 4bbl 325hp—Three or four-speed manual
CJA—402ci V-8 4bbl 325hp—Three or four-speed manual
JK—396ci V-8 4bbl 325hp—Turbo-Hydramatic automatic
CJK—402ci V-8 4bbl 325hp—Turbo-Hydramatic automatic
JC—396ci V-8 4bbl 350hp—Three or four-speed manual
CJC—402ci V-8 4bbl 350hp—Three or four-speed manual
JE—396ci V-8 4bbl 350hp—Turbo-Hydramatic automatic
CJE—402ci V-8 4bbl 350hp—Turbo-Hydramatic automatic

JD—396ci V-8 4bbl 375hp—Four-speed manual
CJD—402ci V-8 4bbl 375hp—Four-speed manual
KF—396ci V-8 4bbl 375hp—Turbohydramatic automatic
CKF—402ci V-8 4bbl 375hp—Turbohydramatic automatic
KG—396ci V-8 4bbl 375hp—Four-speed manual (L89)
CKG—402ci V-8 4bbl 375hp—Four-speed manual (L89)
KH—396ci V-8 4bbl 375hp—Turbo-Hydramatic automatic (L89)
CKH—402ci V-8 4bbl 375hp—Turbo-Hydramatic automatic (L89)

C-10, C-20 Series

PA—250ci I-6 1bbl 155hp—three-speed manual
PC—250ci I-6 1bbl 155hp—manual trans
PD—250ci I-6 1bbl 155hp—Turbo-Hydramatic
PE—250ci I-6 1bbl 155hp—manual trans
PF—250ci I-6 1bbl 155hp—three-speed manual
PG—250ci I-6 1bbl 155hp—three speed manual
PH—250ci I-6 1bbl 155hp—Turbo-Hydramatic
PI—250ci I-6 1bbl 155hp—manual trans
PJ—250ci I-6 1bbl 155hp—manual trans
PP—250ci I-6 1bbl 155hp—manual trans/AIR
PQ—250ci I-6 1bbl 155hp—manual trans/AIR
PW—250ci I-6 1bbl 155hp—automatic trans
RC—292ci I-6 1bbl 170hp—Powerglide
RD—292ci I-6 1bbl 170hp—Turbo-Hydramatic
RG—292ci I-6 1bbl 170hp—manual trans
RH—292ci I-6 1bbl 170hp—manual trans/AIR
RJ—292ci I-6 1bbl 170hp—manual trans
RM—292ci I-6 1bbl 170hp—Powerglide/AIR
RN—292ci I-6 1bbl 170hp—Turbo-Hydramatic/AIR
RV—292ci I-6 1bbl 170hp—manual trans/AIR
UA—307ci V-8 2bbl 200hp—manual trans
UC—307ci V-8 2bbl 200hp—Powerglide
UM—307ci V-8 2bbl 200hp—Turbo-Hydramatic
UN—307ci V-8 2bbl 200hp—Turbo-Hydramatic
VB—350ci V-8 4bbl 255hp—manual trans
WH—350ci V-8 4bbl 255hp—manual trans/AIR
WJ—350ci V-8 4bbl 255hp—Powerglide
WK—350ci V-8 4bbl 255hp—Turbo-Hydramatic
WZ—350ci V-8 4bbl 255hp—three-speed
XA—350ci V-8 4bbl 255hp—manual trans
XC—350ci V-8 4bbl 255hp—Powerglide
XD—350ci V-8 4bbl 255hp—Turbo-Hydramatic
XF—350ci V-8 4bbl 255hp—Powerglide
XG—350ci V-8 4bbl 255hp—Turbo-Hydramatic
XP—350ci V-8 4bbl 255hp—manual trans/AIR
ZA—350ci V-8 4bbl 255hp—manual trans/AIR
ZD—350ci V-8 4bbl 255hp—Powerglide
ZF—350ci V-8 4bbl 255hp—Turbo-Hydramatic
YP—396ci V-8 4bbl 310hp—Turbo-Hydramatic
YQ—396ci V-8 4bbl 310hp—Turbo-Hydramatic
YR—396ci V-8 4bbl 310hp—manual trans

C-20 Series

PA—250ci I-6 1bbl 155hp—manual trans
PB—250ci I-6 1bbl 155hp—manual trans
PE—250ci I-6 1bbl 155hp—manual trans
PF—250ci I-6 1bbl 155hp—manual trans
PG—250ci I-6 1bbl 155hp—manual trans

PH—250ci I-6 1bbl 155hp—manual trans
PI—250ci I-6 1bbl 155hp—manual trans
PJ—250ci I-6 1bbl 155hp—manual trans
PL—250ci I-6 1bbl 155hp—four-speed trans
PN—250ci I-6 1bbl 155hp—manual trans
PP—250ci I-6 1bbl 155hp—manual trans/AIR
PQ—250ci I-6 1bbl 155hp—manual trans
PW—250ci I-6 1bbl 155hp—automatic trans
PX—250ci I-6 1bbl 155hp—automatic trans
PY—250ci I-6 1bbl 155hp—Powerglide
PZ—250ci I-6 1bbl 155hp—four-speed manual
QA—250ci I-6 1bbl 155hp—manual trans
RA—292ci I-6 1bbl 170hp—Powerglide
RB—292ci I-6 1bbl 170hp—Turbo-Hydramatic
RC—292ci I-6 1bbl 170hp—Powerglide
RD—292ci I-6 1bbl 170hp—Turbo-Hydramatic
RE—292ci I-6 1bbl 170hp—manual trans
RG—292ci I-6 1bbl 170hp—manual trans/AIR
RH—292ci I-6 1bbl 170hp—manual trans/AIR
RI—292ci I-6 1bbl 170hp—manual trans
RJ—292ci I-6 1bbl 170hp—manual trans
RM—292ci I-6 1bbl 170hp—Powerglide/AIR
RN—292ci I-6 1bbl 170hp—Turbo-Hydramatic/AIR
RV—292ci I-6 1bbl 170hp—manual trans
SJ—292ci I-6 1bbl 170hp—manual trans/AIR
UA—307ci V-8 2bbl 200hp—manual trans/AIR
UB—307ci V-8 2bbl 200hp—manual trans
UC—307ci V-8 2bbl 200hp—Powerglide
UD—307ci V-8 2bbl 200hp—Powerglide/AIR
UE—307ci V-8 2bbl 200hp—Turbo-Hydramatic/AIR
UN—307ci V-8 2bbl 200hp—Turbo-Hydramatic
UU—307ci V-8 2bbl 200hp—manual trans
VB—350ci V-8 4bbl 255hp—manual trans
VR—350ci V-8 4bbl 255hp—three-speed trans
WA—350ci V-8 4bbl 255hp—Turbo-Hydramatic/AIR
WH—350ci V-8 4bbl 255hp—manual trans/AIR
WJ—350ci V-8 4bbl 255hp—Powerglide
WK—350ci V-8 4bbl 255hp—Turbo-Hydramatic
WL—350ci V-8 4bbl 255hp—Powerglide/AIR
WM—350ci V-8 4bbl 255hp—Turbo-Hydramatic/AIR
WR—350ci V-8 4bbl 255hp—Powerglide
WS—350ci V-8 4bbl 255hp—Turbo-Hydramatic
WT—350ci V-8 4bbl 255hp—manual trans
WZ—350ci V-8 4bbl 255hp—three speed manual
XA—350ci V-8 4bbl 255hp—manual trans/AIR
XB—350ci V-8 4bbl 255hp—manual trans
XC—350ci V-8 4bbl 255hp—Powerglide
XD—350ci V-8 4bbl 255hp—Turbo-Hydramatic/AIR
XE—350ci V-8 4bbl 255hp—manual trans
XF—350ci V-8 4bbl 255hp—Powerglide
XG—350ci V-8 4bbl 255hp—Turbo-Hydramatic
XP—350ci V-8 4bbl 255hp—manual trans
XZ—350ci V-8 4bbl 255hp—Powerglide/AIR
ZA—350ci V-8 4bbl 255hp—manual trans
ZB—350ci V-8 4bbl 255hp—manual trans
ZC—350ci V-8 4bbl 255hp—manual trans
ZE—350ci V-8 4bbl 255hp—manual trans
ZF—350ci V-8 4bbl 255hp—Turbo-Hydramatic

YP—396ci V-8 4bbl 310hp—Turbo-Hydramatic
YQ—396ci V-8 4bbl 310hp—Turbo-Hydramatic
YR—396ci V-8 4bbl 310hp—manual trans
YS—396ci V-8 4bbl 310hp—manual trans
ZA—350ci V-8 4bbl 255hp—manual trans

C-30 Series

PI—250ci I-6 1bbl 155hp—manual trans
PJ—250ci I-6 1bbl 155hp—manual trans
PN—250ci I-6 1bbl 155hp—manual trans
PP—250ci I-6 1bbl 155hp—manual trans/AIR
PQ—250ci I-6 1bbl 155hp—manual trans/AIR
PW—250ci I-6 1bbl 155hp—automatic trans/AIR
PX—250ci I-6 1bbl 155hp—automatic trans
PY—250ci I-6 1bbl 155hp—Powerglide
PZ—250ci I-6 1bbl 155hp—four speed trans
RB—292ci I-6 1bbl 170hp—Turbo-Hydramatic
RD—292ci I-6 1bbl 170hp—Turbo-Hydramatic/AIR
RE—292ci I-6 1bbl 170hp—manual trans
RG—292ci I-6 1bbl 170hp—manual trans/AIR
RI—292ci I-6 1bbl 170hp—manual trans
RN—292ci I-6 1bbl 170hp—Turbo-Hydramatic/AIR
RV—292ci I-6 1bbl 170hp—manual trans/AIR
SJ—292ci I-6 1bbl 170hp—manual trans
UE—307ci V-8 2bbl 200hp—Turbo-Hydramatic
UN—307ci V-8 2bbl 200hp—Turbo-Hydramatic
VR—350ci V-8 4bbl 255hp—three-speed manual
XA—350ci V-8 4bbl 255hp—manual trans/AIR
XB—350ci V-8 4bbl 255hp—manual trans
XD—350ci V-8 4bbl 255hp—Turbo-Hydramatic/AIR
WH—350ci V-8 4bbl 255hp—manual trans/AIR
WJ—350ci V-8 4bbl 255hp—Powerglide
WK—350ci V-8 4bbl 255hp—Turbo-Hydramatic/AIR
WM—350ci V-8 4bbl 255hp—Turbo-Hydramatic/AIR
WR—350ci V-8 4bbl 255hp—Powerglide/AIR
WS—350ci V-8 4bbl 255hp—Turbo-Hydramatic/AIR
WT—350ci V-8 4bbl 255hp—manual trans/AIR
XE—350ci V-8 4bbl 255hp—manual trans
XF—350ci V-8 4bbl 255hp—Powerglide
XG—350ci V-8 4bbl 255hp—Turbo-Hydramatic
XP—350ci V-8 4bbl 255hp—manual trans
ZA—350ci V-8 4bbl 255hp—manual trans
ZC—350ci V-8 4bbl 255hp—manual trans
ZF—350ci V-8 4bbl 255hp—manual trans
YP—396ci V-8 4bbl 310hp—Turbo-Hydramatic/AIR
YQ—396ci V-8 4bbl 310hp—Turbo-Hydramatic
YR—396ci V-8 4bbl 310hp—manual trans
YS—396ci V-8 4bbl 310hp—manual trans

K-10 Series

PA—250ci I-6 1bbl 155hp—manual trans
PC—250ci I-6 1bbl 155hp—manual trans/AIR
PE—250ci I-6 1bbl 155hp—manual trans/AIR
PG—250ci I-6 1bbl 155hp—manual trans
PH—250ci I-6 1bbl 155hp—manual trans
TD—250ci I-6 1bbl 155hp—manual trans
PI—292ci I-6 1bbl 170hp—manual trans
PJ—292ci I-6 1bbl 170hp—manual trans

PP—292ci I-6 1bbl 170hp—manual trans
PQ—292ci I-6 1bbl 170hp—manual trans/AIR
PW—292ci I-6 1bbl 170hp—manual trans/AIR
RD—292ci I-6 1bbl 170hp—Turbo-Hydramatic/AIR
RG—292ci I-6 1bbl 170hp—manual trans/AIR
RH—292ci I-6 1bbl 170hp—manual trans/AIR
RJ—292ci I-6 1bbl 170hp—manual trans
RM—292ci I-6 1bbl 170hp—Powerglide/AIR
RN—292ci I-6 1bbl 170hp—Turbo-Hydramatic/AIR
UO—307ci V-8 2bbl 200hp—manual trans
UQ—307ci V-8 2bbl 200hp—Turbo-Hydramatic
UT—307ci V-8 2bbl 200hp—manual trans
XW—350ci V-8 4bbl 255hp—manual trans
XY—350ci V-8 4bbl 255hp—Turbo-Hydramatic
WN—350ci V-8 4bbl 255hp—manual trans/AIR
WP—350ci V-8 4bbl 255hp—Turbo-Hydramatic
ZG—350ci V-8 4bbl 255hp—three-speed manual
ZJ—350ci V-8 4bbl 255hp—Turbo-Hydramatic

K-20 Series

PA—250ci I-6 1bbl 155hp—manual trans
PB—250ci I-6 1bbl 155hp—manual trans
PE—250ci I-6 1bbl 155hp—manual trans/AIR
PG—250ci I-6 1bbl 155hp—manual trans
PH—250ci I-6 1bbl 155hp—manual trans
PI—250ci I-6 1bbl 155hp—manual trans
PJ—250ci I-6 1bbl 155hp—manual trans
PL—250ci I-6 1bbl 155hp—four-speed trans
PN—250ci I-6 1bbl 155hp—manual trans
PP—250ci I-6 1bbl 155hp—manual trans/AIR
PQ—250ci I-6 1bbl 155hp—manual trans/AIR
PW—250ci I-6 1bbl 155hp—Powerglide/AIR
PX—250ci I-6 1bbl 155hp—Powerglide
PY—250ci I-6 1bbl 155hp—automatic trans
PZ—250ci I-6 1bbl 155hp—four-speed
QA—250ci I-6 1bbl 155hp—manual trans
RB—292ci I-6 1bbl 170hp—Turbo-Hydramatic
RD—292ci I-6 1bbl 170hp—Turbo-Hydramatic
RE—292ci I-6 1bbl 170hp—manual trans
RG—292ci I-6 1bbl 170hp—manual trans/AIR
RH—292ci I-6 1bbl 170hp—manual trans/AIR
RJ—292ci I-6 1bbl 170hp—manual trans
RM—292ci I-6 1bbl 170hp—Powerglide/AIR
RN—292ci I-6 1bbl 170hp—Turbo-Hydramatic/AIR
SJ—292ci I-6 1bbl 170hp—manual trans
UF—307ci V-8 2bbl 200hp—Turbo-Hydramatic
UO—307ci V-8 2bbl 200hp—manual trans/AIR
UP—307ci V-8 2bbl 200hp—manual trans
UT—307ci V-8 2bbl 200hp—manual trans
UV—307ci V-8 2bbl 200hp—manual trans
UW—307ci V-8 2bbl 200hp—manual trans
VS—350ci V-8 4bbl 255hp—three-speed
WN—350ci V-8 4bbl 255hp—manual trans
WO—350ci V-8 4bbl 255hp—manual trans
WP—350ci V-8 4bbl 255hp—Turbo-Hydramatic
WQ—350ci V-8 4bbl 255hp—Turbo-Hydramatic/AIR
XO—350ci V-8 4bbl 255hp—Turbo-Hydramatic/AIR
XW—350ci V-8 4bbl 255hp—manual trans/AIR

XX—350ci V-8 4bbl 255hp—manual trans
XY—350ci V-8 4bbl 255hp—Turbo-Hydramatic
ZG—350ci V-8 4bbl 255hp—three-speed
ZH—350ci V-8 4bbl 255hp—three-speed
ZJ—350ci V-8 4bbl 255hp—Turbo-Hydramatic

Transmission Codes

Code	Type	Plant
B	TH350	Cleveland
C	Powerglide	Cleveland
H	3 speed HD	Muncie
K	3 speed	McKinnon
M	3 speed	Muncie
N	4 speed	Muncie
O	3 speed OD	Saginaw
P	4 speed	Muncie
R	4 speed	Saginaw
S	3 speed	Saginaw
T	Powerglide	Toledo
Y	TH350	Toledo

Axle Identification

Code	Ratio
El Camino	
CL, CM, CM, CO	
CQ, CS	2.56:1
CH, CP, CT	
KC, KD, KE	2.73:1
CD, CX, KK,	3.07:1
CA, CE, KG, KH	3.08:1
CF, CW	3.31:1
CB, CG, KI, KL	3.36:1
KA, KB, KF, KJ	
KN, KP	3.55:1
CI	3.73:1
KK	4.10:1
KM	4.56:1
KO	4.88:1

C-10 Series

HB, LL, RA, RT,	
RU, RW	
TR, TT, WH	3.07:1
HA, HC, TQ, TU	3.73:1
HR, JQ, TS, TV	4.11:1

C-20 Series

NA, NB, OT, OW	
OY, TF	4.10:1
HW, JV	4.11:1
OX, OZ	4.56:1
HU, HV, HX, HY	
JS, OU	4.57:1

C-30 Series

IE, LF, LZ, MR	
NZ, RN	4.10:1
IF, IG, LW, LX	
MC, NW, VL, VM	
VN, VO	4.57:1
ID, LV, LY, TX	
TY, VF, VS	5.14:1
MX	6.17:1

K-10 Series

NX	3.07:1
HQ	3.73:1

K-20 Series

HZ, NC, ND, NE	4.57:1

Exterior Color Codes

C-10, C-20, C-30 Series

Black	500
Light Green	503
Yellow Green	504
Dark Green	505
Medium Blue	506
Light Blue	507
Dark Blue	508
Turquoise	511
Red	514
Orange	516
Maroon	517
Dark Yellow	519
Yellow	520

White	521
Silver	523
Saddle	525

El Camino

Tuxedo Black	10
Butternut Yellow	40
Dover White	50
Dusk Blue	51
Garnet Red	52
Glacier Blue	53
Azure Turquoise	55
Fathom Green	57
Frost Green	59
Burnished Brown	61
Champagne	63
Olympic Gold	65
Burgundy	67
Cortez Silver	69
LeMans Blue	71
Monaco Orange	72*
Daytona Yellow	76*

*SS only

Regular Production Options

El Camino

13380	El Camino 6cyl	$2,550
13580	Custom El Camino 6cyl	2,630
13480	El Camino V-8	2,640
13680	Custom El Camino V-8	2,725
13680	El Camino SS	3,293

C-10 Series

CS10703	Chassis and cab	2,320
CS10704	Stepside pickup	2,400
CS10734	Fleetside pickup	2,435
CS10904	Stepside pickup	2,435
CS10905	Panel delivery	2,865
CS10906	Suburban w/rear doors	3,100
CS10934	Fleetside pickup	2,475

C-20 Series

CS20903	Chassis and cab	2,520
CS20904	Stepside pickup	2,625
CS20905	Panel delivery	3,075
CS20906	Suburban w/rear doors	3,350
CS20909	Stake bed	2,702
CS20916	Suburban w/rear gate	3,385

C-30 Series

CS31003	Chassis & cab	2,627
CS31004	Stepside pickup	2,761
CS31009	Stake bed	2,966
CS31034	Stepside pickup	2,822

K-10 Blazer

KS10514	Utility pickup	2,850

El Camino

Option Number

AR1	Less head restraint
AS1	Standard shoulder harness
A01	Tinted glass (all windows)
A02	Tinted glass (windshield)
A31	Electric control windows
A39	Custom deluxe seatbelts
A51	Strato-Bucket seats
A85	Custom deluxe shoulder belts
A93	Vacuum operated door locks
BX4	Bodyside molding
B37	Color keyed floor mats
B90	Side window moldings
B93	Door edge guards
CE1	Headlamp washer
C08	Vinyl roof cover
C24	Special windshield wiper
C50	Rear window defroster
C60	Deluxe A/C
D33	Remote control outside LH mirror
D34	Visor vanity mirror
D55	Console
D96	Wide side paint stripe
F41	Special performance front & rear suspension
GT1	2.56:1 axle ratio
G76	3.36:1 rear axle ratio
G80	Positraction rear axle
G82	4.56:1 axle ratio
G92	3.08:1 rear axle ratio
G94	3.31:1 rear axle ratio
G96	3.55:1 rear axle ratio
G97	2.73:1 rear axle ratio
HO1	3.07:1 rear axle ratio
HO5	3.73:1 rear axle ratio
J50	Vacuum power brake equipment
J52	Front disc brakes
KD5	HD engine ventilation
K02	Temperature controlled radiator fan
K05	Engine block heater
K79	42 amp AC generator
K85	63 amp AC generator
LM1	350ci V-8 engine (regular fuel)
L22	250ci L6 engine
L34	350hp 396ci Hi-Performance V-8 engine

L35	325hp 396ci V-8 engine
L48	300hp 350ci V-8 engine
L78	375hp 396ci V-8 Special Hi-performance V-8 engine
L89	Aluminum cylinder heads
MC1	HD 3 speed transmission
M20	Four speed wide range transmission
M21	Four speed close ratio transmission
M22	Four speed transmission heavy duty (L78 only)
M35	Powerglide transmission
M38	300 Deluxe 3 speed automatic transmission
M40	Turbo-Hydramatic transmission
NC8	Chambered exhaust system
N10	Dual exhaust
N33	Comfortilt steering wheel
N34	Sports styled steering wheel
N40	Power steering
N95	Simulated wire wheel covers
N96	Simulated magnesium wheel trim cover type A
PA2	Simulated magnesium wheel trim cover type B
P01	Wheel trim cover
P06	Wheel trim ring
T50	Heavy duty battery
UF1	Map lamp
U05	Dual horn
U14	Instrument panel gauges
U15	Speed warning indicator
U25	Luggage compartment lamp
U26	Underhood lamp
U27	Glove compartment lamp
U28	Ashtray lamp
U29	Instrument panel courtesy lights
U35	Electric clock
U46	Light monitoring system
U57	Stereo tape system
U63	Radio—push button control
U69	Radio AM/FM push button control
U79	AM/FM stereo radio
V01	HD radiator
V31	Front bumper guards
V32	Rear bumper guards
V75	Traction compound & dispenser
ZJ7	Rally wheels
ZJ9	Auxiliary lighting group
ZK3	Deluxe seatbelts & front seat shoulder harness
Z25	SS equipment

C10, C20, C30 Series

A07	Glass tinted (10 windows)
A08	Tinted RH body side glass (4 windows)
AO9	Laminated glass
A10	Full view rear window
A11	Soft ray tinted glass
A12	Rear window glass
A18	Swing-out rear door glass
A24	Cab corner windows
A34	Bostrom driver's seat
A35	Bostrom passenger seat
A50	Bucket seats
A52	Custom bench seat
A54	Full width front seat
A55	Level ride seat
A56	Bostrom Levelair Bostrom seat
A57	Auxiliary seat
A59	Supplementary seat
A61	Auxiliary seat, stationary
A62	Front seat belt
A63	Less rear seat belt
A78	Center seat
A80	Center and rear seat
A85	Shoulder harness
A94	Door safety lock
A97	Spare wheel lock
A99	Instrument panel compartment
AA2	Tinted windshield glass
AM2	Heavy duty seat
AM3	Front seat center belt
AN2	Level ride driver & auxiliary seat
AN4	Heavy duty driver & auxiliary seat
AS3	Rear seat
AS5	Shoulder harness, center and rear seats
AU2	Cargo door lock unit
B30	Floor and toe panel carpet
B55	Full foam seat cushion
B59	Padded seat back frame
B70	Instrument panel pad
B85	Body side moulding, upper
B93	Door edge guards
B98	Side trim molding
BE2	Padded hinge pillar
BX1	Body side molding, belt
BX2	Body side molding, wide lower
C07	Auxiliary top
C20	Single speed windshield wiper
C41	Heater—economy
C42	Heater Deluxe
C48	Heater delete
C60	Air conditioning, all weather
C69	Air conditioning, roof mounted
C70	Air conditioning

D14	Arm rest front door
D20	Sunshade, windshield, RH
D23	Sunshade, padded LH
D29	West coast mirror—Jr
D30	West coast mirror—Sr
D32	Rear view mirror
D36	Non-glare inside mirror
D48	Front cross view mirror
D89	Body paint stripe
DG4	West coast mirror, Jr. stainless steel
DG5	West coast mirror, Sr, driver & passenger side
DG8	Outside rearview mirror RH (fixed arm)
DG9	Outside rearview mirror, RH (swinging arm)
DH3	Outside rear view mirror LH,RH (swinging arm)
E23	HD cab lifting torsion bar
E28	Assist handles
E56	Platform and stake rack
E57	Platform body
E80	Pickup box mounting brackets
E81	Floor board
E85	Body side door equipment
FO3	Heavy duty frame
F06	Frame reinforcement, outside inverted "L" type (SAE 1023)
F07	Frame reinforcement, outside inverted "L" type (SAE 950)
F08	Frame reinforcement, outside upright "L" type (SAE 950)
F10	Frame reinforcement, outside inverted "L" type (heat treated)
F19	Special body cross sill mounting support
F25	Frame rails, full depth, heat treated
F43	Front axle, 9000#
F44	Front axle, 11000#
F45	Front axle, 15000#
F47	Front axle, 5000#
F48	Front axle, 7000#
F49	Heavy duty four wheel drive front axle (K-20)
F51	Front and rear HD shock absorbers
F59	Front stabilizer bar
F60	Heavy duty front springs
F76	Front wheel locking hubs (Dana)
F82	Front springs, soft ride, tapered leaf, 7000#
F83	Front springs, soft ride, tapered leaf, 9000#

F84	Front springs, soft ride, tapered leaf, 11000#
F87	Front springs HD 7000#
F88	Front springs HD 9000#
F92	Front springs 8000#
F94	Front springs 9000#
F95	Front springs 10500#
G49	Rear springs, 22000#
G50	Heavy rear springs 2000# ea.
G60	Auxiliary rear springs
G68	Rear shock absorbers
G70	Rear suspension—leaf spring
G80	Rear axle, Positraction
G86	No-Spin rear axle
G87	Rear axle, power lock
G94	Rear axle, 3.31:1
H01	Rear axle, 3.07:1
H04	Rear axle, 4.11:1
H05	Rear axle, 3.73:1
H06	Rear axle, 4.11:1
H07	Single speed rear axle, 6.4 ratio, 13500#
H09	Rear axle 4.11:1, 3600#
J50	Vacuum power brake
J56	Heavy duty brake
J65	Metallic brake facing
J66	15x5 rear brakes
J74	Parking brake, spring loaded
J91	Trailer brake equipment
JA1	Rear axle, 4.10:1 ratio (C-20)
JA4	Dual brake system
JP1	Frame mounted brake booster, hydraulic
K02	Fan drive
K05	Engine block heater
K12	Oil filter, 2qt capacity
K19	AIR equipment
K21	Engine controlled combustion system
K24	Closed engine positive engine ventilation
K28	Fuel filter equipment
K29	Vacuum connector
K31	Throttle control, manual
K37	Engine governor
K43	Air cleaner, dry inside type
K47	Air cleaner equipment
K48	Oil bath air cleaner
K56	Air compressor equipment
K66	Transistor ignition
K67	Heavy duty starter motor
K76	61 amp Delcotron generator
K77	55 amp Delcotron generator
K79	42 amp Delcotron generator
K81	62 amp Delcotron generator
K84	47 amp Delcotron generator
L05	130 amp Delcotron generator

L22	Engine 250ci L-6	U10	Voltmeter
L25	Engine 292ci L-6	U16	Tachometer
L47	Engine 396ci V-8	U30	Pressure gauge
M01	Heavy duty clutch	U31	Ammeter
M04	Clutch 13″, 1 plate	U60	Radio—manual
M05	Clutch 13″, 2 plate	U63	Radio—pushbutton
M07	Clutch 14″, 2 plate	U85	Trailer light cable, 6 wire
M13	Transmission, three-speed HD	U86	Trailer jumper cable
M16	Transmission, three-speed HD	U87	Trailer light cable, 7 wire
M20	Transmission, four-speed	U92	HD wiring
M24	Transmission, four-speed HD	U98	Junction box and wiring
M28	Transmission, four-speed HD close ratio	V01	HD radiator
		V04	Radiator shutters
M35	Powerglide transmission	V05	HD cooling
M49	Transmission, Turbo-Hydramatic	V35	Wraparound front bumper
		V37	Custom chrome option
NO1	Fuel tank—21 gallon	V38	Rear bumper—painted
NO2	Fuel tank—30 gallon	V62	Hydraulic jack
N03	Fuel tank, dual, 37 gallon	V66	Provisions for front end drive power take-off
N35	Steering wheel, 22″		
N40	Power steering	V75	Hazard and marker lights
PO1	Wheel trim cover	V76	Front tow hooks
P10	Spare wheel carrier—under frame mounting	X56	Front bumper construction
		X58	Cab insulation
P13	Spare wheel carrier—side mounted	X59	Cab HD insulation
		Z62	Custom comfort and appearance option
T60	Heavy duty battery		
TP2	Auxiliary battery, camper equipped trucks	Z69	Motorhome chassis conversion
		Z70	$^3/_4$ ton special 7800# GVW
U01	Roof marker and identification lamps	Z81	Camper special
		Z84	Custom Sport Truck package
U06	Air horn	ZJ8	Dual rear wheel conversion
U08	Dual horns		

Facts

The 1969 trucks got a new grille—an aluminum bar with Chevrolet lettering connected the two headlight openings. A large bow-tie emblem was located on the center of the hood, just above the grille.

The side molding on the CST-equipped trucks was slightly different; it didn't follow the wheel arches and its center included a wood-grain effect. A total 29,942 trucks were equipped with the CST option.

In the interior, a two-spoke steering wheel replaced the previous three-spoke design. The parking brake was also redesigned, using a foot-operated lever. The dash was updated with new gauges and an optional tachometer.

Two new models were introduced. The Longhorn pickup was an extended Fleetside pickup using the 133in wheelbase. It was available as a C-20 or a C-30. Fleetside pickups could be equipped with either a wooden load floor or an all-steel floor.

Also new was the Blazer. It was basically a shortened Fleetside pickup, as it used a 104in wheelbase and was fitted with a removable fiberglass top. A total 4,935 were sold—4,636 with a white top and 290 with a black top. The Blazer was available only as a four-wheel drive this year and thus came with leaf springs. In CST form, it came with carpets, interior door and box trim panels, front bucket seats and a removable rear bench seat.

Air conditioning was becoming more popular, and 61,568 trucks were so equipped.

A 350ci version of the small-block V-8 replaced the 327. It was equipped with a four-barrel Rochester Quadrajet carburetor for a 255hp output. All other engines were carry-overs from 1968.

The SS 396 was no longer a separate model series. It became an option, RPO Z25, which was available on the El Camino. Because Chevrolet listed only the total quantity of SS 396 options sold, the number that were El Caminos cannot be determined. SS El Caminos came with a blacked-out grille and rear taillight panel. The wheelwells got bright moldings, and for the first time, chrome five-spoke Magnum wheels with SS center caps were standard. SS 396 emblems were located on the grille, the rear panel, and the front fenders.

In the interior, SS emblems were displayed on the steering wheel and the dash.

All 396 El Caminos were equipped with power front disc brakes as standard equipment. If power steering was ordered, variable quick-ratio steering was included.

Engine availability on the SS model was unchanged: the standard L35 325hp 396ci, the L34 350hp 396ci, and the L78 375hp 396ci.

During the model year, the 396's bore was increased from 4.094in to 4.125in, resulting in 402ci. The engine was still, however, marketed as the SS 396 in this and subsequent years.

1969 El Camino.

Chapter 26

1970 Trucks

Production

13380 El Camino 6cyl	1,303	
13580 Custom El Camino 6cyl	539	
13480 El Camino V-8	5,137	
13680 Custom El Camino V-8	40,728	
Total El Camino	47,707	

C-10 Series		K Series
CS10703 Chassis and cab	2,084	64
CS10704 Stepside pickup	31,353	1,629
CS10734 Fleetside pickup	40,754	2,554
CS10903 Chassis and cab	913	26
CS10904 Stepside pickup	11,857	464
CS10905 Panel delivery	3,965	—
CS10906 Suburban w/rear doors	5,927	926
CS10934 Fleetside pickup	234,904	7,348

C-20 Series		
CS20903 Chassis and cab	7,277	582
CS20904 Stepside pickup	5,856	953
CS20905 Panel delivery	1,032	—
CS20906 Suburban w/rear doors	1,344	287
CS20916 Suburban w/rear gate	880	254

C-30 Series		
CS31003 Chassis & cab	14,873	
CS31004 Stepside pickup	2,101	
CS31034 Stepside pickup	1,404	

K-5 Blazer		
CS10514 Utility pickup	985	11,527

Serial numbers

Description

CE107040A100001
C—Chassis type, C—conventional, K—four-wheel drive
E—Engine: E—V-8, S—V-6
1—GVW range: 1—3900-5800lb, 2—5200-7500lb, 3—6600-14,000lb
07—Cab to axle dimension: 05—30-35″, 09—54-59″, 10—60-65″, 14—84-89″
04—Body type, 02—Cowl, 03—Cab, 04—Stepside pickup, 05—Panel,
    06—Suburban (panel rear doors), 09—Platform stake, 12—Windshield cowl,
    14—Utility Blazer, 16—Suburban (tail & liftgate), 34—Fleetside pickup
0—Last digit of model year, 0—1970

A—Assembly Plant Code: A—Atlanta GA, B—Baltimore MD, F—Flint MI,
   J—Janesville WI, S—St. Louis MO, T—Tarrytown NY, Z—Fremont CA,
   1—Oshawa Canada
100001—Consecutive Sequence Number

Location: On plate attached to left door hinge post. On cowl models, plate is attached to left cowl inner panel.

El Camino Serial Number

136800A100001
1—Chevrolet Division
36—Series, 33—El Camino 6cyl, 34—El Camino 8cyl, 35—Custom El Camino
   6cyl, 36—Custom El Camino 8cyl, 38—El Camino SS
80—Body style, 2 door sedan pickup
0—Last digit of model year, 0—1970
A—Assembly Plant Code: A—Atlanta GA, B—Baltimore MD, F—Flint MI,
   K—Kansas City MO, L—Van Nuys CA, 1—Oshawa Canada
100001—Consecutive Sequence Number

Location: On plate attached to driver's side of dash, visible through the windshield.

Model & Wheelbase

Model Number and Description

Model Number and Description	Wheelbase (in)
El Camino	
13380 El Camino 6cyl	115
13480 El Camino 8cyl	115
13580 El Camino Custom 6cyl	115
13680 El Camino Custom 8cyl	115
C-5 Series	
CE10514 Blazer	104
C-10 Series	
CE10702 Chassis and cowl	115
CE10703 Chassis and cab	115
CE10704 Stepside pickup	115
CE10734 Fleetside pickup	115
CE10903 Chassis & cab	127
CE10904 Stepside pickup	127
CE10905 Panel delivery	127
CE10906 Suburban w/rear doors	127
CE10916 Suburban w/tailgate	127
CE10934 Fleetside pickup	127
C-20 Series	
CE20902 Chassis and cowl	127
CE20903 Chassis and cab	127
CE20904 Stepside pickup	127
CE20905 Panel delivery	127
CE20906 Suburban w/rear doors	127
CE20909 Stake bed	127
CE20912 Windshield Cowl	127

| CE20916 | Suburban w/tailgate | 127 |
| CE20934 | Longhorn pickup | 133 |

C-30 Series
CE31002	Chassis & cowl	133
CE31003	Chassis & cab	133
CE31004	Stepside pickup	133
CE31009	Stake bed	133
CE31403	Chassis & cab	157

K-5 Series
| KE10514 | Blazer | 104 |

K-10 Series
KE10703	Chassis and cab	115
KE10704	Stepside pickup	115
KE10734	Fleetside pickup	115
KE10903	Chassis & cab	127
KE10904	Stepside pickup	127
KE10905	Panel delivery	115
KE10906	Suburban w/rear doors	127
KE10916	Suburban w/tailgate	127
KE10934	Fleetside pickup	127

K-20 Series
KE20903	Chassis and cab	127
KE20904	Stepside pickup	127
KE20905	Panel delivery	127
KE20906	Suburban w/rear doors	127
KE20916	Suburban w/tailgate	127
KE20934	Fleetside pickup	127

Engine & Transmission Suffix Codes

El Camino
CRG—250ci I-6 1bbl 155hp—manual trans
CCH—250ci I-6 1bbl 155hp—Powerglide
CCG—250ci I-6 1bbl 155hp—automatic trans
CCF—250ci I-6 1bbl 155hp—automatic trans
CCM—250ci I-6 1bbl 155hp—Powerglide
CCK—250ci I-6 1bbl 155hp—Turbo-Hydramatic
CCL—250ci I-6 1bbl 155hp—manual trans
CRF—250ci I-6 1bbl 155hp—manual trans
CNC—307ci V-8 2bbl 200hp—manual trans
CND—307ci V-8 2bbl 200hp—manual trans
CNE—307ci V-8 2bbl 200hp—Powerglide
CNF—307ci V-8 2bbl 200hp—Turbo-Hydramatic
CNG—307ci V-8 2bbl 200hp—manual trans
CNH—307ci V-8 2bbl 200hp—Powerglide
CNI—350ci V-8 2bbl 250hp—manual trans
CNJ—350ci V-8 4bbl 300hp—manual trans
CNK—350ci V-8 4bbl 300hp—Powerglide
CNN—350ci V-8 2bbl 250hp—Turb-Hydramatic
CRE—350ci V-8 4bbl 300hp—Turbo-Hydramatic
CNM—350ci V-8 4bbl 300hp—Powerglide
CZX—402ci V-8 2bbl 265hp—manual trans
CRH—402ci V-8 2bbl 265hp—Turbo-Hydramatic

155

CKN—402ci V-8 4bbl 330hp—Four-speed manual
CKR—402ci V-8 4bbl 330hp—Four-speed manual
CKS—402ci V-8 4bbl 350hp—Four-speed manual
CTX—402ci V-8 4bbl 350hp—Four-speed manual
CTW—402ci V-8 4bbl 350hp—Turbo-Hydramatic
CKO—402ci V-8 4bbl 375hp—Four-speed manual
CTY—402ci V-8 4bbl 375hp—Turbo-Hydramatic
CKT—402ci V-8 4bbl 375hp—Four-speed manual, aluminum heads
CKP—402ci V-8 4bbl 375hp—Turbo-Hydramatic, aluminum heads
CRT—454ci V-8 4bbl 360hp—Four-speed manual
CRQ—454ci V-8 4bbl 360hp—Turbo-Hydramatic
CRV—454ci V-8 4bbl 450hp—Four-speed manual
CRR—454ci V-8 4bbl 450hp—Turbo-Hydramatic

C-10 Series
TAA—250ci I-6 1bbl 155hp—four-speed manual
TAB—250ci I-6 1bbl 155hp—manual trans
TAC—250ci I-6 1bbl 155hp—automatic trans
TAD—250ci I-6 1bbl 155hp—Turbo-Hydramatic
TCE—250ci I-6 1bbl 155hp—manual trans
TCF—250ci I-6 1bbl 155hp—automatic trans
TCG—250ci I-6 1bbl 155hp—Turbo-Hydramatic
TCH—250ci I-6 1bbl 155hp—three-speed manual
TCJ—250ci I-6 1bbl 155hp—Powerglide
TCK—250ci I-6 1bbl 155hp—four-speed manual
TCL—250ci I-6 1bbl 155hp—manual trans
TCU—292ci I-6 1bbl 170hp—Powerglide
TCZ—292ci I-6 1bbl 170hp—manual trans
TNA—292ci I-6 1bbl 170hp—manual trans
TNB—292ci I-6 1bbl 170hp—manual trans
TNC—292ci I-6 1bbl 170hp—Powerglide
TAF—292ci I-6 1bbl 170hp—Turbo-Hydramatic
TND—292ci I-6 1bbl 170hp—manual trans
TAI—307ci V-8 2bbl 200hp—Powerglide
TAJ—307ci V-8 2bbl 200hp—Powerglide
TAK—307ci V-8 2bbl 200hp—Turbo-Hydramatic
TAS—307ci V-8 2bbl 200hp—manual trans
TAU—350ci V-8 4bbl 255hp—manual trans
TAV—350ci V-8 4bbl 255hp—manual trans
TAZ—350ci V-8 4bbl 255hp—Powerglide
TAX—350ci V-8 4bbl 255hp—Turbo-Hydramatic
TBA—350ci V-8 4bbl 255hp—Turbo-Hydramatic
TMJ—350ci V-8 4bbl 255hp—Turbo-Hydramatic
TBB—350ci V-8 4bbl 255hp—manual trans
TBD—350ci V-8 4bbl 255hp—manual trans
TNT—350ci V-8 4bbl 255hp—Powerglide
TBG—402ci V-8 4bbl 310hp—Turbo-Hydramatic
TBH—402ci V-8 4bbl 310hp—manual trans

C-20 Series
TAA—250ci I-6 1bbl 155hp—four-speed manual
TAB—250ci I-6 1bbl 155hp—three-speed manual
TAC—250ci I-6 1bbl 155hp—automatic trans
TAD—250ci I-6 1bbl 155hp—Turbo-Hydramatic
TCE—250ci I-6 1bbl 155hp—three-speed manual
TCF—250ci I-6 1bbl 155hp—automatic trans
TCG—250ci I-6 1bbl 155hp—Turbo-Hydramatic
TCH—250ci I-6 1bbl 155hp—three-speed manual

TCJ—250ci I-6 1bbl 155hp—Powerglide
TCK—250ci I-6 1bbl 155hp—four-speed manual
TCL—250ci I-6 1bbl 155hp—manual trans
TCU—292ci I-6 1bbl 170hp—Powerglide
TCZ—292ci I-6 1bbl 170hp—manual trans
TNA—292ci I-6 1bbl 170hp—manual trans
TNB—292ci I-6 1bbl 170hp—manual trans
TNC—292ci I-6 1bbl 170hp—Powerglide
TND—292ci I-6 1bbl 170hp—manual trans
TAF—292ci I-6 1bbl 170hp—Turbo-Hydramatic
TAM—307ci V-8 2bbl 200hp—manual trans
TAR—307ci V-8 2bbl 200hp—manual trans
TAH—307ci V-8 2bbl 200hp—Turbo-Hydramatic
TAJ—307ci V-8 2bbl 200hp—Powerglide
TAV—350ci V-8 4bbl 255hp—three-speed manual
TAZ—350ci V-8 4bbl 255hp—Powerglide
TBB—350ci V-8 4bbl 255hp—three-speed manual
TBD—350ci V-8 4bbl 255hp—manual trans
TMJ—350ci V-8 4bbl 255hp—Turbo-Hydramatic
TNT—350ci V-8 4bbl 255hp—Powerglide
TBG—402ci V-8 4bbl 310hp—Turbo-Hydramatic
TBH—402ci V-8 4bbl 310hp—manual trans

C-30 Series
TAA—250ci I-6 1bbl 155hp—four-speed manual
TCF—250ci I-6 1bbl 155hp—automatic trans
TCK—250ci I-6 1bbl 155hp—manual trans
TCL—250ci I-6 1bbl 155hp—manual trans
TAF—292ci I-6 1bbl 170hp—Turbo-Hydramatic
TCZ—292ci I-6 1bbl 170hp—manual trans
TNA—292ci I-6 1bbl 170hp—manual trans
TND—292ci I-6 1bbl 170hp—manual trans
TNE—292ci I-6 1bbl 170hp—Turbo-Hydramatic
TAL—307ci V-8 2bbl 200hp—Turbo-Hydramatic
TAM—307ci V-8 2bbl 200hp—manual trans
TAR—307ci V-8 2bbl 200hp—manual trans
TAV—350ci V-8 4bbl 255hp—three-speed manual
TAX—350ci V-8 4bbl 255hp—Turbo-Hydramatic
TBA—350ci V-8 4bbl 255hp—Turbo-Hydramatic
TBB—350ci V-8 4bbl 255hp—three-speed manual
TMJ—350ci V-8 4bbl 255hp—Turbo-Hydramatic
TBD—350ci V-8 4bbl 255hp—manual trans
TBG—402ci V-8 4bbl 310hp—Turbo-Hydramatic
TBH—402ci V-8 4bbl 310hp—manual trans

K-10 Series
TAB—250ci I-6 1bbl 155hp—three-speed manual
TAD—250ci I-6 1bbl 155hp—Turbo-Hydramatic
TCH—250ci I-6 1bbl 155hp—three-speed manual
TCK—250ci I-6 1bbl 155hp—four-speed manual
TCL—250ci I-6 1bbl 155hp—manual trans
TNA—292ci I-6 1bbl 170hp—manual trans
TCZ—292ci I-6 1bbl 170hp—manual trans
TNB—292ci I-6 1bbl 170hp—manual trans
TAF—292ci I-6 1bbl 170hp—Turbo-Hydramatic
TAO—307ci V-8 2bbl 200hp—Turbo-Hydramatic
TAT—307ci V-8 2bbl 200hp—three-speed manual
TBE—307ci V-8 2bbl 200hp—three-speed manual

TAY—350ci V-8 4bbl 255hp—Turbo-Hydramatic
TBC—350ci V-8 4bbl 255hp—three-speed manual
TBF—350ci V-8 4bbl 255hp—three-speed manual

K-20 Series
TAB—250ci I-6 1bbl 155hp—three-speed manual
TAD—250ci I-6 1bbl 155hp—Turbo-Hydramatic
TCH—250ci I-6 1bbl 155hp—three-speed manual
TCK—250ci I-6 1bbl 155hp—four-speed manual
TCL—250ci I-6 1bbl 155hp—manual trans
TNA—292ci I-6 1bbl 170hp—manual trans
TCZ—292ci I-6 1bbl 170hp—manual trans
TNB—292ci I-6 1bbl 170hp—manual trans
TAF—292ci I-6 1bbl 170hp—Turbo-Hydramatic
TAN—307ci V-8 2bbl 200hp—three-speed manual
TAP—307ci V-8 2bbl 200hp—Turbo-Hydramatic
TAW—307ci V-8 2bbl 200hp—three-speed manual
TAY—307ci V-8 2bbl 200hp—Turbo-Hydramatic
TBC—350ci V-8 4bbl 255hp—three-speed manual
TBF—350ci V-8 4bbl 255hp—three-speed manual

Transmission Codes

Code	Type	Plant
A	Torque Drive	Cleveland*
B	TH	Cleveland
C	Powerglide	Cleveland
K	3 speed	McKinnon
M	3 speed	Muncie
N	4 speed	Muncie
O	3 speed OD	Saginaw
P	4 speed	Muncie
R	4 speed	Saginaw
S	3 speed	Saginaw
Y	TH	Toledo

*El Camino

Axle Identification

Code	Ratio
El Camino	
CCL, CCM, CCN, CCO	
CGA, CGB, CRJ, CRK	2.56:1
CCH, CCP, CGC, CGD,	
CKC, CKD	2.73:1
WC, XC	2.78:1
CCD, CCX, CRL, CRM	3.07:1
CCA, CCE, CGE, CGF	
WE, XE	3.08:1
WF, XF	3.23:1
CCF, CCW, CRU, CRV	3.31:1
CCB, CGG, CGI	3.36:1
CKF, CKJ	3.55:1
CKK, CRW	4.10:1

C-10 Series

TDC, TDD, TDE, TDI	
TPH, TPI, TPJ, TPL	
TPO, TPP	3.07:1
TDA, TDH, TPF, TPM	3.73:1
TDB, TDJ, TPG, TPN	4.11:1

C-20 Series

TAA, THU, THW, TJU,	
TLN, TMZ, TNA, TNB,	
TOT, TTF, TWI	4.10:1
TYS	4.56:1
THV, THX, THY, TIA,	
TIB, TJS, TJT, TOU,	4.57:1

C-30 Series

TIE, TIH, TLJ, TLX,	
TLZ, TMR, TNZ, TRN,	4.10:1
TIG, TLW, TMC, TNW,	
TVL, TVM, TVN, TVO	4.57:1
TID, TIF, TLV, TLY,	
TVF, TVI, TVS, TVT	5.14:1
TMX	6.17:1

K-10 Series

TDG, TDI, TPL, TPO,	
TPP, TPR	3.07:1
TDF, TDH, TPK	3.73:1
TPW, TPX	4.10:1
TDJ, TPN,	4.11:1

K-20 Series

TPW, TPX	4.10:1
TGA, TGB, TPZ	4.57:1

Exterior Color Codes

C-10, C-20, C-30 Series

Black	500
Medium Blue	501
Yellow Green	504
Dark Green	505
Dark Olive	506
Dark Blue	508
Medium Blue	510
Turquoise	511
Dark Blue Green	512
Flame Red	513
Red	514
Orange	516
Medium Green	518
Yellow	519
Light Yellow	520
White	521
Copper	522
Red-Orange	524
Medium Gold	526
Dark Gold	527
Dark Bronze	528
Medium Blue-Green	529
Dark Blue	570

El Camino

Classic White	10
Cortez Silver	14
Shadow Gray	17
Tuxedo Black	19
Astro Blue	25
Fathom Blue	28
Misty Turquoise	34
Green Mist	45
Forest Green	48
Gobi Beige	50
Champagne Gold	55
Autumn Gold	58
Desert Sand	63
Cranberry Red	75
Black Cherry	78

Regular Production Options

El Camino

13380	El Camino 6cyl	$2,679
13580	Custom El Camino 6cyl	2,769
13480	El Camino V-8	2,770
13680	Custom El Camino V-8	2,850

C-10 Series

CS10504	Blazer (4x2)	2,385
CS10703	Chassis and cab	2,405
CS10704	Stepside pickup	2,520
CS10734	Fleetside pickup	2,558
CS10904	Stepside pickup	2,560
CS10905	Panel delivery	3,071
CS10906	Suburban w/rear doors	3,250
CS10934	Fleetside pickup	2,595

Add $570 for 115″ & 127″ WB 4WD models

C-20 Series

CS20903	Chassis and cab	2,652
CS20904	Stepside pickup	2,752
CS20905	Panel delivery	3,270
CS20906	Suburban w/rear doors	3,440
CS20916	Suburban w/rear gate	3,385

Add $705 for 127″ WB 4WD models

C-30 Series

CS31003	Chassis & cab	2,736
CS31004	Stepside pickup	2,874
CS31009	Stake bed	3,075
CS31034	Stepside pickup	2,935

K-5 Blazer

KS10514	Utility pickup	2,956

El Camino

Option Number

AK1	Deluxe seatbelts & front shoulder harness
AQ2	Electric seatbelt lock release
AU3	Electric door locks
A01	Tinted glass (all windows)
A02	Tinted glass (windshield)
A31	Electric control windows
A39	Custom deluxe seatbelts
A41	4-way electric control front seat
A46	4-way electric control front bucket seat
A51	Strato-Bucket seats
A85	Deluxe shoulder harness
B37	Color keyed floor mats
B85	Belt reveal molding
B90	Side window moldings
B93	Door edge guards
CD2	Windshield washer fluid level
CD3	Electro tip windshield wiper
C08	Exterior soft trim roof cover
C50	Rear window defroster
C60	Deluxe A/C
D33	Remote control outside LH mirror

D34	Visor vanity mirror	
D55	Console	
D88	Sport stripe	
F40	Special front & rear suspension	
F41	Special performance front & rear suspension	
G67	Rear shock absorber level control	
G80	Positraction rear axle	
J50	Vacuum power brake equipment	
K05	Engine block heater	
K30	Speed & cruise control	
K85	63 amp AC generator	
LF6	400ci 2bbl V-8 engine	
LS3	400ci V-8 engine	
LS6	454ci Special Hi-Performance V-8 engine	
L48	350ci V-8 engine	
L65	350ci 2bbl V-8 engine	
L78	396ci V-8 Special Hi-performance V-8 engine	
L89	Aluminum cylinder heads	
M20	Four speed wide range transmission	
M21	Four speed close ratio transmission	
M22	Four speed transmission heavy duty	
M35	Powerglide transmission	
M38	300 Deluxe 3 speed automatic transmission	
M40	Turbo-Hydramatic transmission	
NA9	EEC	
NK1	Cushioned rim steering wheel	
N10	Dual exhaust	
N33	Comfortilt steering wheel	
N40	Power steering	
PA3	Special wheel trim cover	
P01	Wheel trim cover	
P06	Wheel trim ring	
T58	Rear wheel opening skirt	
T60	Heavy duty battery	
UM1	Push-button AM radio & tape player	
UM2	Push-button AM/FM stereo radio & tape player	
U14	Instrument panel gauges	
U35	Electric clock	
U46	Lamp monitoring system	
U57	Stereo tape system	
U63	Radio—push button control	
U69	Radio AM/FM push button control	
U76	Windshield antenna	
U79	AM/FM stereo radio	
V01	HD radiator	

V31	Front bumper guards
V32	Rear bumper guards
YD1	Axle for trailering
ZJ7	Special wheel, hubcap & trim ring
ZJ9	Auxiliary lighting group
ZL2	Special ducted hood air system
ZQ9	Performance ratio rear axle
Z15	SS 454 equipment
Z25	SS 396 equipment

C10, C20, C30 Series

A07	Glass tinted (10 windows)
A08	Tinted RH body side glass (4 windows)
AO9	Laminated glass
A10	Full view rear window
A11	Soft ray tinted glass
A12	Rear window glass
A18	Swing-out rear door glass
A24	Cab corner windows
A34	Bostrom driver's seat
A35	Bostrom passenger seat
A50	Bucket seats
A52	Custom bench seat
A54	Full width front seat
A55	Level ride seat
A56	Bostrom Levelair Bostrom seat
A57	Auxiliary seat
A59	Supplementary seat
A61	Auxiliary seat, stationary
A62	Front seat belt
A63	Less rear seat belt
A78	Center seat
A80	Center and rear seat
A85	Shoulder harness
A94	Door safety lock
A97	Spare wheel lock
A99	Instrument panel compartment
AA2	Tinted windshield glass
AM2	Heavy duty seat
AM3	Front seat center belt
AN2	Level ride driver & auxiliary seat
AN4	Heavy duty driver & auxiliary seat
AS3	Rear seat
AS5	Shoulder harness, center and rear seats
AU2	Cargo door lock unit
B30	Floor and toe panel carpet
B55	Full foam seat cushion
B59	Padded seat back frame
B70	Instrument panel pad
B85	Body side moulding, upper
B93	Door edge guards
B98	Side trim molding
BE2	Padded hinge pillar

BX1	Body side molding, belt
BX2	Body side molding, wide lower
C07	Auxiliary top
C20	Single speed windshield wiper
C41	Heater—economy
C42	Heater Deluxe
C48	Heater delete
C60	Air conditioning, all weather
C69	Air conditioning, roof mounted
C70	Air conditioning
D14	Arm rest front door
D20	Sunshade, windshield, RH
D23	Sunshade, padded LH
D29	West coast mirror—Jr
D30	West coast mirror—Sr
D32	Rear view mirror
D36	Non-glare inside mirror
D48	Front cross view mirror
D89	Body paint stripe
DG4	West coast mirror, Jr. stainless steel
DG5	West coast mirror, Sr, driver & passenger side
DG8	Outside rearview mirror RH (fixed arm)
DG9	Outside rearview mirror, RH (swinging arm)
DH3	Outside rear view mirror LH,RH (swinging arm)
E23	HD cab lifting torsion bar
E28	Assist handles
E30	Series P10 body equipment
E31	Series P20-30 body equipment
E32	Series P20-30 body equipment
E33	Series P20-P30 body equipment (aluminum)
E56	Platform and stake rack
E57	Platform body
E80	Pickup box mounting brackets
E81	Floor board
E85	Body side door equipment
F02	Special heavy duty frame (24000# GVW on 60 series)
FO3	Heavy duty frame
F19	Special body cross sill mounting support
F25	Frame rails, full depth, heat treated
F43	Front axle, 9000#
F44	Front axle, 11000#
F45	Front axle, 15000#
F47	Front axle, 5000#
F48	Front axle, 7000#
F49	Heavy duty four wheel drive front axle (K-20)
F51	Front and rear HD shock absorbers
F54	Front axle, 12000#

F59	Front stabilizer bar
F60	Heavy duty front springs
F76	Front wheel locking hubs (Dana)
F82	Front springs, soft ride, tapered leaf, 7000#
F83	Front springs, soft ride, tapered leaf, 9000#
F84	Front springs, soft ride, tapered leaf, 11000#
F87	Front springs HD 7000#
F88	Front springs HD 9000#
F91	Front springs 13000#
F92	Front springs 8000#
F94	Front springs 9000#
F95	Front springs 10500#
F96	Front springs 11000#
F98	Front springs 14000#
G50	Heavy rear springs 2000# ea.
G60	Auxiliary rear springs
G68	Rear shock absorbers
G70	Rear suspension—leaf spring
G80	Rear axle, Positraction
G86	No-Spin rear axle
G87	Rear axle, power lock
G94	Rear axle, 3.31:1
H01	Rear axle, 3.07:1
H04	Rear axle, 4.11:1
H05	Rear axle, 3.73:1
H06	Rear axle, 4.11:1
H09	Rear axle 4.11:1, 3600#
J56	Heavy duty brake
J65	Metallic brake facing
J66	15x5 rear brakes
J70	Brake booster, hydraulic
J73	Heavy duty brake booster, hydraulic
J74	Parking brake, spring loaded
J91	Trailer brake equipment
JA1	Rear axle, 4.10:1 ratio (C-20)
JA4	Dual brake system
JP1	Frame mounted brake booster, hydraulic
K02	Fan drive
K05	Engine block heater
K12	Oil filter, 2qt capacity
K19	AIR equipment
K21	Engine controlled combustion system
K24	Closed engine positive engine ventilation
K28	Fuel filter equipment
K29	Vacuum connector
K31	Throttle control, manual
K37	Engine governor
K43	Air cleaner, dry inside type
K47	Air cleaner equipment
K48	Oil bath air cleaner

K56	Air compressor equipment
K67	Heavy duty starter motor
K76	61 amp Delcotron generator
K77	55 amp Delcotron generator
K79	42 amp Delcotron generator
K81	62 amp Delcotron generator
K84	47 amp Delcotron generator
L05	130 amp Delcotron generator
L22	Engine 250ci L-6
L25	Engine 292ci L-6
M01	Heavy duty clutch
M13	Transmission, three-speed HD
M16	Transmission, three-speed HD
M20	Transmission, four-speed
M24	Transmission, four-speed HD
M28	Transmission, four-speed HD close ratio
M35	Powerglide transmission
M49	Transmission, Turbo-Hydramatic
N03	Fuel tank, dual, 37 gallon
N35	Steering wheel, 22"
N40	Power steering
PO1	Wheel trim cover
P10	Spare wheel carrier—under frame mounting
P13	Spare wheel carrier—side mounted
T60	Heavy duty battery
TP2	Auxiliary battery, camper equipped trucks
U01	Roof marker and identification lamps
U06	Air horn
U08	Dual horns
U10	Voltmeter
U16	Tachometer
U30	Pressure gauge
U31	Ammeter
U60	Radio—manual
U63	Radio—pushbutton
U85	Trailer light cable, 6 wire
U86	Trailer jumper cable
U87	Trailer light cable, 7 wire
U92	HD wiring
U98	Junction box and wiring
V01	HD radiator
V04	Radiator shutters
V05	HD cooling
V35	Wraparound front bumper
V37	Custom chrome option
V38	Rear bumper—painted
V62	Hydraulic jack
V66	Provisions for front end drive power take-off
V75	Hazard and marker lights
V76	Front tow hooks
X56	Front bumper construction
X58	Cab insulation
X59	Cab HD insulation
Z62	Custom Comfort and appearance option
Z69	Motorhome chassis conversion
Z70	¾ ton special 7800# GVW
Z81	Camper special
Z84	Custom Sport Truck
ZJ8	Dual rear wheel conversion

Facts

The 1970 trucks got a new grille that included a series of six horizontal ribs above and below the bright centerpiece. Otherwise, the trucks were visually unchanged.

A total 49,717 trucks came with the Custom Port Sports Truck Package.

The 396ci big-block V-8, which had undergone a bore increase, was now officially known as the 400, even though it displaced 402ci. It should not be confused with the 400ci version of the small-block V-8. The small-block V-8 was known as the Turbo-Fire, whereas the big-block was a Turbo-Jet.

This was the last year of the panel model.

Considered the high point in performance, the 1970 El Camino got new front and rear end treatments. The El Camino did not have side vent windows. As the El Camino SS option was included with the Chevelle, it is not known how many El Camino SSs were built.

Two SS packages were available. Z25 was the SS 396 option, and Z15 was the SS 454 option package. Both packages included a blacked-out grille; a domed hood; wheelwell moldings; a black ream insert panel on the rear insert panel bumper; insert panel, chrome 14x7 five-spoke Sport wheels with F70x14

RWL raised-white-letter tires; twin rectangular exhaust outlets; an SS grille emblem; an SS rear bumper emblem; SS emblems on the black steering wheel and column; and, unique to SS models, a black-faced dash panel that came with round gauges.

SS 396 or SS 454 emblems were included and mounted on the front fenders. Mechanically, the SS came with the F41 suspension, power front disc brakes, and either the 402ci engine (marketed as a 396) or the 454ci V-8. The standard L34 402ci was rated at 350hp, and the optional L78 was rated at 375hp. Only eighteen aluminum-head L89s were built, as the option was deleted early in the model year.

The big deal was, of course, the LS5 and LS6 454ci engines. Featuring a larger bore and stroke than the 402—4.25x4—the 454 represented the largest production permutation of the Mark IV big-block series. The LS5, rated at 360hp, came with the small oval port cylinder heads, a Quadrajet carburetor, a cast-iron intake manifold, and a hydraulic camshaft. Available with either the four-speed or the Turbo Hydra-matic automatic, an LS5-equipped Chevelle could be optioned with air conditioning. The LS6, rated at 450hp, came with the large rectangular port heads, an aluminum high-rise intake and Holley carburetor, a solid-lifter camshaft, and a one-point-higher compression ratio, 11.25:1.

Optional on the SS cars was the cowl induction hood. It was a domed hood that came with a vacuum-operated rear-facing air valve on the back of the domed hood that fed outside air into the engine's air cleaner. Cowl Induction lettering was used on both sides, and black or white "Band-Aid" stripes on the hood and deck were part of the package. The stripes were also available without the cowl induction option. Hood lock pins were included with the Cowl Induction Package.

Conventional antennas were eliminated on all El Caminos. The radio antenna was now a wire imbedded in the front windshield. SS Chevelles came with clear, rather than amber, front turn signal lamps.

An additional 402ci big-block engine was available on all other, non-SS El Caminos. The engine was known as the Turbo-Jet 400, and cars so equipped could be identified by 400 emblems on the front fenders.

1971 Chevrolet "Cheyenne" Fleetside Pickup. Chevrolet Motor Division

Chapter 27

1971 Trucks

Production

Vega
14105 Panel Express	7,819

El Camino
13380 El Camino 6 cyl	1,058
13480 El Camino V-8	4,506
13680 El Camino Custom V-8	36,042
Total El Camino	41,606

C-5 Series
		K-Series
CS10514 Utility pickup	1,277	17,220

C-10 Series
		K-Series
CS10703 Chassis and cab	1,476	
CS10704 Stepside pickup	19,041	1,438
CS10734 Fleetside pickup	32,865	3,068
CS10903 Chassis and cab	588	—
CS10904 Stepside pickup	7,269	364
CS10906 Suburban w/rear doors	4,550	631
CS10934 Fleetside pickup	206,313	9,417

C-20 Series
CS20903 Chassis and cab	4,523	509
CS20904 Stepside pickup	3,523	674
CS20906 Suburban w/rear doors	1,343	256
CS20916 Suburban w/rear gate	1,203	353
CS20934 Fleetside pickup	62,465	10,006
CS21034 Longhorn pickup	3,331	—

C-30 Series
CS31003 Chassis & cab	11,438
CS31004 Stepside pickup	1,557
CS31034 Stepside pickup	1,479

Serial numbers

Description
CE107041A100001
C—Chassis type, C—conventional, K—four-wheel drive
E—Engine: E—V-8, S—V-6
1—GVW range: 1—3900-5800lb, 2—5200-7500lb, 3—6600-14,000lb
07—Cab to axle dimension: 05—30-35″, 09—54-59″, 10—60-65″, 14—84-89″

04—Body type, 02—Cowl, 03—Cab, 04—Stepside pickup, 05—Panel, 06—
   Suburban (panel rear doors), 09—Platform stake, 12—Windshield cowl, 14—
   Utility Blazer, 16—Suburban (tail & liftgate), 34—Fleetside pickup
1—Last digit of model year, 1—1971
A—Assembly Plant Code: A—Lakewood GA, B—Baltimore MD, F—Flint MI,
   G—Flint MI, J—Janesville WI, K—Leeds MO, L—Van Nuys CA, P—Pontiac
   MI, R—Arlington TX, S—St. Louis MO, T—Tarrytown NY, Z—Fremont CA,
   1—Oshawa, Canada
100001—Consecutive Sequence Number

Location: On plate attached to left door hinge post. On cowl models, plate
is attached to left cowl inner panel.

El Camino Serial Number

136801B100001
1—Chevrolet Division
36—Series, 33—El Camino 6cyl, 34—El Camino 8cyl, 36—Custom El Camino
   8cyl
80—Body style, 2 door sedan pickup
1—Last digit of model year, 1—1971
B—Assembly Plant Code: B—Baltimore MD, K—Kansas City MO, L—Van
   Nuys CA, R—Arlington TX
100001—Consecutive Sequence Number

Location: On left plate attached to driver's side of dash, visible through
the windshield.

Model & Wheelbase

Model Number and Description	Wheelbase (in)
Vega	
14105 Panel Express	97
El Camino	
13380 El Camino 6cyl	116
13480 El Camino 8cyl	116
13680 El Camino Custom 8cyl	116
C-5 Series	
CE10514 Blazer	104
C-10 Series	
CE10702 Chassis and cowl	115
CE10703 Chassis and cab	115
CE10704 Stepside pickup	115
CE10734 Fleetside pickup	115
CE10903 Chassis & cab	127
CE10904 Stepside pickup	127
CE10906 Suburban w/rear doors	127
CE10916 Suburban w/tailgate	127
CE10934 Fleetside pickup	127
C-20 Series	
CE20902 Chassis and cowl	127
CE20903 Chassis and cab	127

CE20904	Stepside pickup	127
CE20906	Suburban w/rear doors	127
CE20909	Stake bed	127
CE20912	Windshield cowl	127
CE20916	Suburban w/tailgate	127
CE21034	Longhorn pickup	133

C-30 Series

CE31002	Chassis & cowl	133
CE31003	Chassis & cab	133
CE31004	Stepside pickup	133
CE31009	Stake bed	133
CE31034	Longhorn pickup	133
CE31403	Chassis & cab	157

K-5 Series

| KE10514 | Blazer | 104 |

K-10 Series

KE10703	Chassis and cab	115
KE10704	Stepside pickup	115
KE10734	Fleetside pickup	115
KE10903	Chassis & cab	127
KE10904	Stepside pickup	127
KE10906	Suburban w/rear doors	127
KE10916	Suburban w/tailgate	127
KE10934	Fleetside pickup	127

K-20 Series

KE20903	Chassis and cab	127
KE20904	Stepside pickup	127
KE20906	Suburban w/rear doors	127
KE20916	Suburban w/tailgate	127
KE20934	Fleetside pickup	127

Engine & Transmission Suffix Codes

El Camino

CAG—250ci I-6 1bbl 145hp—manual trans
CAB—250ci I-6 1bbl 145hp—Powerglide
CAA—250ci I-6 1bbl 145hp—manual trans
CCA—307ci V-8 2bbl 200hp—Powerglide
CCC—307ci V-8 2bbl 200hp—manual trans
CGA—350ci V-8 2bbl 245hp—manual trans
CGB—350ci V-8 2bbl 245hp—Powerglide
CGK—350ci V-8 4bbl 270hp—manual trans
CGL—350ci V-8 2bbl 270hp—Turbo-Hydramatic
CJD—350ci V-8 4bbl 270hp—Turbo-Hydramatic
CJJ—350ci V-8 4bbl 270hp—manual trans
CGC—350ci V-8 2bbl 245hp—Turbo-Hydramatic
CLK—402ci V-8 2bbl 255hp—Turbo-Hydramatic
CAJ—402ci V-8 2bbl 255hp—manual trans
CLL—402ci V-8 4bbl 300hp—manual trans
CLS—402ci V-8 4bbl 300hp—three speed manual, HD
CLA—402ci V-8 4bbl 300hp—manual trans
CLB—402ci V-8 4bbl 300hp—Turbo-Hydramatic

CPA—454ci V-8 4bbl 365hp—four-speed manual
CPD—454ci V-8 4bbl 365hp—Turbo-Hydramatic

C-10 Series
TCD—250ci I-6 1bbl 145hp—manual trans
TCS—250ci I-6 1bbl 145hp—automatic trans
TCT—250ci I-6 1bbl 145hp—Turbo-Hydramatic
TCP—250ci I-6 1bbl 145hp—manual trans
TGA—292ci I-6 1bbl 165hp—manual trans
TGB—292ci I-6 1bbl 165hp—Turbo-Hydramatic
TGC—292ci I-6 1bbl 165hp—manual trans
TGD—292ci I-6 1bbl 165hp—Turbo-Hydramatic
TGG—292ci I-6 1bbl 165hp—Turbo-Hydramatic
THA—307ci V-8 2bbl 200hp—manual trans
THC—307ci V-8 2bbl 200hp—Turbo-Hydramatic
THK—307ci V-8 2bbl 200hp—Turbo-Hydramatic
THL—307ci V-8 2bbl 200hp—Turbo-Hydramatic
TBA—350ci V-8 4bbl 250hp—manual trans
TBD—350ci V-8 4bbl 250hp—manual trans
TBC—350ci V-8 4bbl 250hp—Turbo-Hydramatic
TBG—350ci V-8 4bbl 250hp—Turbo-Hydramatic
TBH—350ci V-8 4bbl 250hp—manual trans
TBK—350ci V-8 4bbl 250hp—Turbo-Hydramatic
TBJ—350ci V-8 4bbl 250hp—Powerglide
TKA—402ci V-8 4bbl 300hp—manual trans
TKB—402ci V-8 4bbl 300hp—Turbo-Hydramatic

C-20 Series
TCD—250ci I-6 1bbl 145hp—manual trans
TCS—250ci I-6 1bbl 145hp—automatic trans
TCT—250ci I-6 1bbl 145hp—Turbo-Hydramatic
TCP—250ci I-6 1bbl 145hp—manual trans
TGA—292ci I-6 1bbl 165hp—manual trans
TGB—292ci I-6 1bbl 165hp—Turbo-Hydramatic
TGC—292ci I-6 1bbl 165hp—manual trans
TGD—292ci I-6 1bbl 165hp—Turbo-Hydramatic
TGG—292ci I-6 1bbl 165hp—Turbo-Hydramatic
THK—307ci V-8 2bbl 200hp—Turbo-Hydramatic
THL—307ci V-8 2bbl 200hp—Turbo-Hydramatic
THP—307ci V-8 2bbl 200hp—manual trans
TBA—350ci V-8 4bbl 250hp—manual trans
TBD—350ci V-8 4bbl 250hp—manual trans
TBC—350ci V-8 4bbl 250hp—Turbo-Hydramatic
TBG—350ci V-8 4bbl 250hp—Turbo-Hydramatic
TBH—350ci V-8 4bbl 250hp—manual trans
TBK—350ci V-8 4bbl 250hp—Turbo-Hydramatic
TBJ—350ci V-8 4bbl 250hp—Powerglide
TKA—402ci V-8 4bbl 300hp—manual trans
TKB—402ci V-8 4bbl 300hp—Turbo-Hydramatic

C-30 Series
TCL—250ci I-6 1bbl 145hp—manual trans
TCW—250ci I-6 1bbl 145hp—manual trans
TCX—250ci I-6 1bbl 145hp—automatic trans
TGA—292ci I-6 1bbl 165hp—manual trans
TGB—292ci I-6 1bbl 165hp—Turbo-Hydramatic
TGC—292ci I-6 1bbl 165hp—manual trans
TGD—292ci I-6 1bbl 165hp—Turbo-Hydramatic

TGG—292ci I-6 1bbl 165hp—Turbo-Hydramatic
THP—307ci V-8 2bbl 200hp—manual trans
TBA—350ci V-8 4bbl 250hp—manual trans
TBD—350ci V-8 4bbl 250hp—manual trans
TBC—350ci V-8 4bbl 250hp—Turbo-Hydramatic
TBG—350ci V-8 4bbl 250hp—Turbo-Hydramatic
TBH—350ci V-8 4bbl 250hp—manual trans
TBK—350ci V-8 4bbl 250hp—Turbo-Hydramatic
TBJ—350ci V-8 4bbl 250hp—Powerglide
TKA—402ci V-8 4bbl 300hp—manual trans
TKB—402ci V-8 4bbl 300hp—Turbo-Hydramatic

K-10 Series

TCD—250ci I-6 1bbl 145hp—manual trans
TCS—250ci I-6 1bbl 145hp—automatic trans
TCT—250ci I-6 1bbl 145hp—Turbo-Hydramatic
TGA—292ci I-6 1bbl 165hp—manual trans
TGB—292ci I-6 1bbl 165hp—Turbo-Hydramatic
TGC—292ci I-6 1bbl 165hp—manual trans
TGD—292ci I-6 1bbl 165hp—Turbo-Hydramatic
TGG—292ci I-6 1bbl 165hp—Turbo-Hydramatic
THD—307ci V-8 2bbl 200hp—manual trans
THK—307ci V-8 2bbl 200hp—Turbo-Hydramatic
THL—307ci V-8 2bbl 200hp—Turbo-Hydramatic
TBA—350ci V-8 4bbl 250hp—manual trans
TBD—350ci V-8 4bbl 250hp—manual trans
TBC—350ci V-8 4bbl 250hp—Turbo-Hydramatic
TBG—350ci V-8 4bbl 250hp—Turbo-Hydramatic
TBH—350ci V-8 4bbl 250hp—manual trans
TBK—350ci V-8 4bbl 250hp—Turbo-Hydramatic
TBJ—350ci V-8 4bbl 250hp—Powerglide

K-20 Series

TCD—250ci I-6 145hp—manual trans
TCS—250ci I-6 1bbl 145hp—automatic trans
TCT—250ci I-6 1bbl 145hp—Turbo-Hydramatic
TGA—292ci I-6 1bbl 165hp—manual trans
TGB—292ci I-6 1bbl 165hp—Turbo-Hydramatic
TGC—292ci I-6 1bbl 165hp—manual trans
TGD—292ci I-6 1bbl 165hp—Turbo-Hydramatic
TGG—292ci I-6 1bbl 165hp—Turbo-Hydramatic
THK—307ci V-8 2bbl 200hp—Turbo-Hydramatic
THR—307ci V-8 2bbl 200hp—manual trans
THS—307ci V-8 2bbl 200hp—Turbo-Hydramatic
TBA—350ci V-8 4bbl 250hp—manual trans
TBD—350ci V-8 4bbl 250hp—manual trans
TBC—350ci V-8 4bbl 250hp—Turbo-Hydramatic
TBG—350ci V-8 4bbl 250hp—Turbo-Hydramatic
TBH—350ci V-8 4bbl 250hp—manual trans
TBK—350ci V-8 4bbl 250hp—Turbo-Hydramatic
TBJ—350ci V-8 4bbl 250hp—Powerglide

Transmission Codes

Code	Type	Plant
A	Torque Drive	Cleveland*
B	TH	Cleveland

C	Powerglide	Cleveland
H	3 speed	Muncie*
J	TH	GM/CAN
K	3 speed	McKinnon
M	3 speed	Muncie
N	4 speed	Muncie
P	4 speed	Muncie
R	4 speed	Saginaw
S	3 speed	Muncie
Y	TH	Toledo

*El Camino

Axle Identification

Code	Ratio

El Camino

CB, GA	2.56:1
CH, GC, GD, GH, KD,	2.73:1
GF, GN	3.08:1
CF, CW, RU, RV	3.31:1
GG, GI,	3.36:1
RW	4.10:1

C-10 Series

THA, THB, THS.THT,	
TRR, TRS, TRT, TRW	3.07:1
TAA, THC, THW, THX	3.73:1
THD, THG, THY, THZ	4.11:1

C-20 Series

TAH, TAJ, TJA, TJB,	
TJP, TJS, 2G	4.10:1
TRA, TRB, TRC, TRD,	
TRG, TRH, TRJ	4.56:1
TAB, TGA, THP, TJC,	
TJR, TJT, 2J	4.57:1

C-30 Series

TAK, TAL, TAP, TJK,	4.10:1
TAC, TAD, TAG, TGK,	
THR,	4.57:1
TAR, TJL, TRP, TRZ	5.14:1
TRY	5.43:1
TAS, TAT	6.17:1

K-10 Series

TAY, TAZ, THH, THJ	3.07:1
TAW, TAX, THK, THL	3.73:1

K-20 Series

TJX, TJY, TKA, TKD	
TKG, TKH, TKJ, TKP	
TKR,	3.54:1
TJH, TJJ, TKB, TKC	
TKK, TKL, TKS, TKT	4.10:1
TKY, TKZ	4.11:1

TKW, TKX	4.56:1
TJD, TJG, TJW,	4.57:1

Exterior Color Codes

C-10, C-20, C-30 Series

Black	500
Medium Blue	501
Medium Olive	504
Dark Green	505
Dark Olive	506
Dark Blue	508
Medium Blue	510
Ochre	511
Dark Blue Green	512
Flame Red	513
Red	514
Orange	516
Medium Green	518
Dark Yellow	519
Light Yellow	520
White	521
Medium Bronze	522
Dark Blue	523
Red-Orange	524
Yellow	525

El Camino

Antique White	11
Nevada Silver	13
Tuxedo Black	19
Ascot Blue	24
Mulsanne Blue	26
Cottonwood Green	42
Lime Green	43
Antique Green	49
Sunflower	52
Placer Gold	53
Sandalwood	61
Burnt Orange	62
Classic Copper	67
Cranberry Red	75
Rosewood	78

Regular Production Options

Vega

Panel Express	$2,286

El Camino

13380 El Camino 6cyl.	2,886
13480 El Camino V-8	2,983
13680 Custom El Camino V-8	3,069

C-10 Series

CS10703 Chassis and cab	2,656
CS10704 Stepside pickup	2,816
CS10734 Fleetside pickup	2,930
CS10904 Stepside pickup	2,854
CS10906 Suburban w/rear doors	3,709
CS10934 Fleetside pickup	2,854

Add $570 for 115" & 127" WB 4WD models

C-20 Series

CS20903 Chassis and cab	3,011
CS20904 Stepside pickup	3,058
CS20906 Surburban w/rear doors	3,760
CS20916 Suburban w/rear gate	3,791

Add $680 for 127" WB 4WD models

C-30 Series

CS31003 Chassis & cab	3,102
CS31004 Stepside pickup	3,255
CS31009 Stake bed	3,530
CS31034 Stepside pickup	3,316

K-5 Blazer

KS10514 Utility pickup	3,358

El Camino

Option Number

AK1	Deluxe seatbelts & front shoulder harness
AQ2	Electric seatbelt lock release
AU3	Electric door locks
A01	Tinted glass (all windows)
A02	Tinted glass (windshield)
A31	Electric windows
A39	Custom deluxe seatbelts
A41	4-way electric control front seat
A46	4-way electric control front bucket seat
A51	Strato-Bucket seats
A85	Deluxe shoulder harness
B37	Floor mats
B85	Belt reveal molding
B90	Side window moldings
B93	Door edge guards
C08	Exterior soft trim roof cover
C50	Rear window defrester
C60	Deluxe A/C
D33	Remote control outside LH mirror
D34	Visor vanity mirror
D55	Console
D88	Sport stripe
F40	Special front & rear suspension
F41	Special performance front & rear suspension
G67	Rear shock absorber level control
G80	Positraction rear axle
J50	Vacuum power brake equipment
JL2	Front disc brakes
K05	Engine block heater
K30	Speed control, cruise master
K85	63 amp AC generator
LS3	400ci (402) V-8 engine
LS5	454ci Hi-Performance V-8 engine
L48	350ci V-8 engine
MC1	HD 3 speed transmission
M11	Floor shift transmission
M20	Four speed wide range transmission
M22	Four speed transmission heavy duty
M35	Powerglide transmission
M38	300 Deluxe 3 speed automatic transmission
M40	Turbo-Hydramatic transmission
NK2	Deluxe steering wheel
NK4	Deluxe steering wheel
N33	Tilt-type steering wheel
N40	Power steering
PA3	Special wheel trim cover
P01	Wheel trim cover
P02	Deluxe wheel trim cover
T58	Rear wheel opening skirt
T60	Heavy duty battery
UM1	Push-button AM radio & tape player
UM2	Push-button AM/FM stereo radio & tape player
U14	Instrument panel gauges
U35	Electric clock
U63	Radio—push button control
U69	Radio AM/FM push button control

U76	Windshield antenna	B93	Door edge guards
U79	AM/FM stereo radio	B98	Side trim molding
V01	HD radiator	BE2	Padded hinge pillar
V30	Front & rear bumper guards	BX1	Body side molding, belt
YD1	Axle for trailering	BX2	Body side molding, wide lower
YF3	Heavy Chevy package	C07	Auxiliary top
ZJ7	Special wheel, hubcap & trim ring	C20	Single speed windshield wiper
ZJ9	Auxiliary lighting group	C41	Heater—economy
ZL2	Special ducted hood air system	C42	Heater Deluxe
ZQ9	Performance ratio rear axle	C48	Heater delete
Z15	SS equipment	C60	Air conditioning, all weather
		C69	Air conditioning, roof mounted
		C70	Air conditioning

C10, C20, C30 Series

A07	Glass tinted (10 windows)	D14	Arm rest front door
A08	Tinted RH body side glass (4 windows)	D20	Sunshade, windshield, RH
AO9	Laminated glass	D23	Sunshade, padded LH
A10	Full view rear window	D29	West coast mirror—Jr
A11	Soft ray tinted glass	D30	West coast mirror—Sr
A12	Rear window glass	D32	Rear view mirror
A18	Swing-out rear door glass	D36	Non-glare inside mirror
A24	Cab corner windows	D48	Front cross view mirror
A28	Sliding rear window	D89	Body paint stripe
A34	Bostrom driver's seat	DF1	Mirrors, exterior, painted, camper style
A35	Bostrom passenger seat	DF2	Mirrors, exterior, stainless steel, camper style
A50	Bucket seats		
A52	Custom bench seat	DG4	West coast mirror, Jr. stainless steel
A54	Full width front seat		
A55	Level ride seat	DG5	West coast mirror, Sr, driver & passenger side
A56	Bostrom Levelair Bostrom seat		
A57	Auxiliary seat	DG8	Outside rearview mirror RH (fixed arm)
A59	Supplementary seat		
A61	Auxiliary seat, stationary	DG9	Outside rearview mirror, RH (swinging arm)
A62	Front seat belt		
A63	Less rear seat belt	DH3	Outside rear view mirror LH, RH (swinging arm)
A78	Center seat		
A80	Center and rear seat	E23	HD cab lifting torsion bar
A85	Shoulder harness	E28	Assist handles
A94	Door safety lock	E56	Platform and stake rack
A97	Spare wheel lock	E57	Platform body
A99	Instrument panel compartment	E80	Pickup box mounting brackets
AA2	Tinted windshield glass	E81	Floor board
AM2	Heavy duty seat	E85	Body side door equipment
AM3	Front seat center belt	FO3	Heavy duty frame
AN2	Level ride driver & auxiliary seat	F19	Special body cross sill mounting support
AN4	Heavy duty driver & auxiliary seat	F25	Frame rails, full depth, heat treated
AS3	Rear seat	F43	Front axle, 9000#
AS5	Shoulder harness, center and rear seats	F44	Front axle, 11000#
		F45	Front axle, 15000#
AU2	Cargo door lock unit	F47	Front axle, 5000#
B30	Floor and toe panel carpet	F48	Front axle, 7000#
B55	Full foam seat cushion	F49	Heavy duty four wheel drive front axle (K-20)
B59	Padded seat back frame		
B70	Instrument panel pad	F51	Front and rear HD shock absorbers
B85	Body side moulding, upper		
		F54	Front axle, 12000#

F59	Front stabilizer bar
F60	Heavy duty front springs
F76	Front wheel locking hubs (Dana)
F82	Front springs, soft ride, tapered leaf, 7000#
F83	Front springs, soft ride, tapered leaf, 9000#
F84	Front springs, soft ride, tapered leaf, 11000#
F87	Front springs HD 7000#
F88	Front springs HD 9000#
F92	Front springs 8000#
F94	Front springs 9000#
F95	Front springs 10500#
F96	Front springs 11000#
G50	Heavy rear springs 2000# ea.
G60	Auxiliary rear springs
G64	Rear spring equipment, 34500#
G68	Rear shock absorbers
G70	Rear suspension—leaf spring
G80	Rear axle, Positraction
G86	No-Spin rear axle
G87	Rear axle, power lock
G94	Rear axle, 3.31:1
H01	Rear axle, 3.07:1
H04	Rear axle, 4.11:1
H05	Rear axle, 3.73:1
H20	Rear axle, 4.57:1
H22	Rear axle, 6.17:1
J74	Parking brake, spring loaded
J91	Trailer brake equipment
JA1	Rear axle, 4.10:1 ratio (C-20)
JA4	Dual brake system
JP1	Frame mounted brake booster, hydraulic
K02	Fan drive
K05	Engine block heater
K12	Oil filter, 2qt capacity
K21	Engine controlled combustion system
K24	Closed engine positive engine ventilation
K28	Fuel filter equipment
K29	Vacuum connector
K31	Throttle control, manual
K37	Engine governor
K43	Air cleaner, dry inside type
K47	Air cleaner equipment
K48	Oil bath air cleaner
K56	Air compressor equipment
K66	Transistor ignition
K67	Heavy duty starter motor
K76	61 amp Delcotron generator
K77	55 amp Delcotron generator
K79	42 amp Delcotron generator
K81	62 amp Delcotron generator
K84	47 amp Delcotron generator

LS9	350ci V-8 engine
L05	130 amp Delcotron generator
L22	Engine 250ci L-6
L25	Engine 292ci L-6
L47	Engine 400 (402)ci V-8
M01	Heavy duty clutch
M13	Transmission, three-speed HD
M16	Transmission, three-speed HD
M20	Transmission, four-speed
M24	Transmission, four-speed HD
M28	Transmission, four-speed HD close ratio
M35	Powerglide transmission
M49	Transmission, Turbo-Hydramatic
NO2	Fuel tank—30 gallon
N03	Fuel tank, dual, 37 gallon
N40	Power steering
PO1	Wheel trim cover
P10	Spare wheel carrier—under frame mounting
P13	Spare wheel carrier—side mounted
T60	Heavy duty battery
TP2	Auxiliary battery, camper equipped trucks
U01	Roof marker and identification lamps
U06	Air horn
U08	Dual horns
U10	Voltmeter
U16	Tachometer
U30	Pressure gauge
U31	Ammeter
U60	Radio—manual
U63	Radio—pushbutton
U85	Trailer light cable, 6 wire
U86	Trailer jumper cable
U87	Trailer light cable, 7 wire
U92	HD wiring
U98	Junction box and wiring
V01	HD radiator
V04	Radiator shutters
V05	HD cooling
V35	Wraparound front bumper
V37	Custom chrome option
V38	Rear bumper—painted
V62	Hydraulic jack
V66	Provisions for front end drive power take-off
V75	Hazard and marker lights
V76	Front tow hooks
X56	Front bumper construction
X58	Cab insulation
X59	Cab HD insulation
Z58	Auxiliary top
Z52	Custom comfort and appearance option

Z69   Motorhome chassis conversion
Z70   ³/₄ ton special 7800# GVW
Z81   Camper special
Z84   Custom Sport Truck
ZJ8   Dual rear wheel conversion

Facts

The light trucks got a new egg-crate grille in 1971. The turn signal indicator lamps were relocated in the bumper. The bow-tie emblem, previously located mounted on the hood, was now relocated moved to the center of the grille.

Mechanically, disc brakes became standard equipment on all trucks. These were power assisted on all models except the 1/2-tons, for which power assist was optional. All pickups were equipped with a small yellow caution sticker on the tailgate.

All engines were modified to run on low-lead gasoline. A new interior trim package was available: the Cheyenne Super. It included carpets, vinyl seats and interior trim, chrome bumpers, and special exterior identification. A total 9,867 trucks were so equipped. Additional interior colors were available for the first time as well, such as Dark Saddle, Parchment, and Argent Silver.

Fleetside trucks with optional two-tone paint could be painted ordered with the area between the side moldings painted white. The previously used two-tone scheme with the lower body area painted the second color was also available.

An AM/FM radio was optionally available for the first time, with 5,282 installations.

The production of CST-equipped models rose to 72,609 units.

The El Camino got a new grille with single 7in headlights. In the rear, two round taillight lamps were located in each side of the bumper. On the SS model, SS emblems were located displayed in the center of the front grille and in the center of the rear bumper. A nice addition to the SS was 15x7 gray mag-style steel wheels; the domed hood was still standard. The interior features were the same, including an SS emblem on the black steering wheel.

Mechanically, the El Camino was unchanged, save in the engine compartment. Only one 454ci engine was still available: the LS5, uprated to 365hp, even though compression ratio dropped to 9:1.

Besides the LS5 454ci V-8, three other engines were available on the El Camino SS. First was the L65 245hp 350ci small-block V-8. It came with a 2-barrel carburetor and a single exhaust. Next was the L48 350ci rated at 270hp. Third was the LS3 402ci big-block rated at 300hp. The LS3 was marketed as the Turbo-Jet 400.

Cowl induction was available. The optional power steering included a variable-ratio steering box.

The newest addition to the truck line was the Vega-based Panel Express. Using the Kammback or station wagon body, it came with a standard 90hp 140ci four-cylinder engine, three-speed manual transmission, and front disc brakes as standard equipment. The vinyl interior was available in either black or green. In standard form, the Vega Panel Express came only with a driver's seat. Most regular Vega production options were available.

Chapter 28

1972 Trucks

Production

El Camino

13380 El Camino 6cyl	1,307	
13480 El Camino 8cyl	5,481	
13680 El Camino Custom 8cyl	50,359	
Total El Camino	57,147	

Vega

14150 Panel Expres	4,114	

Luv

82 Mini pickup	21,098	

C-5 Series

		K Series
CS10514 Blazer	3,357	44,266

C-10 Series

CS10703 Chassis and cab	1,640	—
CS10704 Stepside pickup	22,042	1,736
CS10734 Fleetside pickup	39,730	6,069
CS10904 Stepside pickup	7,538	407
CS10906 Suburban w/doors	6,748	991
CS10906 Suburban w/gate	10,757	2,145
CS10934 Fleetside pickup	273,249	18,431

C-20 Series

CS20903 Chassis and cab	5,974	676
CS20904 Stepside pickup	3,973	755
CS20906 Suburban w/doors	2,136	503
CS20916 Suburban w/gate	3,141	879
CE21024 Longhorn pickup	3,328	—

C-30 Series

CE31002 Chassis & cowl	127	
CE31003 Chassis & cab	14,988	
CS31004 Stepside pickup	1,542	
CS31034 Longhorn pickup	2,450	
CE31403 Chassis & cab	8,944	

Serial numbers
Description
CE107042A100001
C—Chassis type, C—conventional, K—four-wheel drive

E—Engine: E—V-8, S—V-6
1—GVW range: 1—3900-5800lb, 2—5200-7500lb, 3—6600-14,000lb
07—Cab to axle dimension: 05—30-35", 09—54-59", 10—60-65", 14—84-89"
04—Body type, 02—Cowl, 03—Cab, 04—Stepside pickup, 06—Suburban
   (panel rear doors), 09—Platform stake, 12—Windshield cowl, 14—Utility
   Blazer, 16—Suburban (tail & liftgate), 34—Fleetside pickup
2—Last digit of model year, 2—1972
A—Assembly Plant Code: A—Lakewood GA, B—Baltimore MD, F—Flint MI,
   J—Janesville WI, K—Leeds MO, L—Van Nuys CA, P—Pontiac MI,
   R—Arlington TX, S—St. Louis MO, T—Tarrytown NY, U—Lordstown OH,
   Z—Fremont CA, 1—Oshawa Canada
100001—Consecutive Sequence Number

Location: On plate attached to left door hinge post. On cowl models, plate
is attached to left cowl inner panel.

El Camino Serial Number
1D80W2B100001
1—Chevrolet
D—Chevelle Malibu V-8
80—body style, 80—2dr sedan pickup
W—Engine code, D—250 L6, F—307 V8, H—350-2 V8, J—350-4 V8, U—402-4
   V8, W—454-4 V8
2—Last digit of model year, 2—1972
A—Assembly Plant Code: B—Baltimore MD, K—Leeds MO, L—Van Nuys CA,
   R—Arlington TX, 1—Oshawa Canada
100001—Consecutive Sequence Number (500001 Oshawa)

Engine Codes
D—250ci 1bbl L-6 110hp
F—307ci 2bbl V-8 130hp
H—350ci 2bbl V-8 165hp

J—350ci 4bbl V-8 175hp
U—402ci 4bbl V-8 240hp
W—454ci 4bbl V-8 270hp

Location: On plate attached to driver's side of dash, visible through the
windshield.

Model & Wheelbase

Model Number and Description | Wheelbase (in)

Model Number and Description	Wheelbase (in)
Luv	
82 Mini pickup	102.4
Vega	
14150 Panel Express	97
El Camino	
13380 El Camino 6cyl	116
13480 El Camino 8cyl	116
13680 El Camino Custom 8cyl	116
C-5 Series	
CE10514 Blazer	104

C-10 Series

CE10702	Chassis and cowl	115
CE10703	Chassis and cab	115
CE10704	Stepside pickup	115
CE10734	Fleetside pickup	115
CE10903	Chassis & cab	127
CE10904	Stepside pickup	127
CE10906	Suburban w/rear doors	127
CE10916	Suburban w/tailgate	127
CE10934	Fleetside pickup	127

C-20 Series

CE20902	Chassis and cowl	127
CE20903	Chassis and cab	127
CE20904	Stepside pickup	127
CE20906	Suburban w/rear doors	127
CE20909	Stake bed	127
CE20912	Windshield Cowl	127
CE20916	Suburban w/tailgate	127
CE20934	Longhorn pickup	133

C-30 Series

CE31002	Chassis & cowl	133
CE31003	Chassis & cab	133
CE31004	Stepside pickup	133
CE31009	Stake bed	133
CE31034	Longhorn pickup	133
CE31403	Chassis & cab	157

K-5 Series

| KE10514 | Blazer | 104 |

K-10 Series

KE10703	Chassis and cab	115
KE10704	Stepside pickup	115
KE10734	Fleetside pickup	115
KE10903	Chassis & cab	127
KE10904	Stepside pickup	127
KE10906	Suburban w/rear doors	127
KE10916	Suburban w/tailgate	127
KE10934	Fleetside pickup	127

K-20 Series

KE20903	Chassis and cab	127
KE20904	Stepside pickup	127
KE20906	Suburban w/rear doors	127
KE20916	Suburban w/tailgate	127
KE20934	Fleetside pickup	127

Engine & Transmission Suffix Codes

El Camino

CBJ—250ci I-6 1bbl 110hp—Powerglide
CBG—250ci I-6 1bbl 110hp—manual trans
CSD—250ci I-6 1bbl 110hp—Powerglide w/NB2
CBA—250ci I-6 1bbl 110hp—manual trans w/NB2

CBK—250ci I-6 1bbl 110hp—Powerglide
CBD—250ci I-6 1bbl 110hp—manual trans w/NB2
CKG—307ci V-8 2bbl 130hp—manual transmission
CAY—307ci V-8 2bbl 130hp—manual transmission w/NB2
CKH—307ci V-8 2bbl 130hp—Powerglide automatic
CAZ—307ci V-8 2bbl 130hp—Powerglide automatic w/NB2
CTK—307ci V-8 2bbl 130hp—Turbo-Hydramatic
CMA—307ci V-8 2bbl 130hp—Turbo-Hydramatic w/NB2
CKA—350ci V-8 2bbl 165hp—manual transmission
CDA—350ci V-8 2bbl 165hp—manual transmission w/NB2
CTL—350ci V-8 2bbl 165hp—Turbo-Hydramatic
CMD—350ci V-8 2bbl 165hp—Turbo-Hydramatic w/NB2
CKK—350ci V-8 4bbl 175hp—manual transmission
CDG—350ci V-8 4bbl 175hp—manual transmission w/NB2
CKD—350ci V-8 4bbl 175hp—Turbo-Hydramatic
CDD—350ci V-8 4bbl 175hp—Turbo-Hydramatic w/NB2
CLA, CLS—402ci V-8 4bbl 240hp—manual transmission
CTA—402ci V-8 4bbl 240hp—manual transmission
CTH—402ci V-8 4bbl 240hp—three-speed HD w/AIR
CLB—402ci V-8 4bbl 240hp—Turbo-Hydramatic
CTB—402ci V-8 4bbl 240hp—Turbo-Hydramatic w/AIR
CPA—454ci V-8 4bbl 270hp—manual transmission
CRX—454ci V-8 4bbl 270hp—manual transmission w/AIR
CRN—454ci V-8 4bbl 270hp—Turbo-Hydramatic
CRW—454ci V-8 4bbl 270hp—Turbo-Hydramatic w/AIR

C-10 Series

TPL—250ci I-6 1bbl 110hp—manual trans
TPY—250ci I-6 1bbl 110hp—Turbo-Hydramatic
TPT—250ci I-6 1bbl 110hp—manual trans
TPW—250ci I-6 1bbl 110hp—manual trans
TPX—250ci I-6 1bbl 110hp—manual trans
TPS—250ci I-6 1bbl 110hp—Turbo-Hydramatic
TPR—250ci I-6 1bbl 110hp—manual trans
THJ—250ci I-6 1bbl 110hp—manual trans
TJL—250ci I-6 1bbl 110hp—manual trans
TLY—250ci I-6 1bbl 110hp—manual trans
TLB—307ci V-8 2bbl 135hp—manual trans
TDA—307ci V-8 2bbl 135hp—manual trans
TDL—307ci V-8 2bbl 135hp—Turbo-Hydramatic
TRA—307ci V-8 2bbl 135hp—manual trans
TAH—307ci V-8 2bbl 135hp—Turbo-Hydramatic
TSP—307ci V-8 2bbl 135hp—Turbo-Hydramatic
TSS—307ci V-8 2bbl 135hp—manual trans
TBL—350ci V-8 4bbl 175hp—manual trans
TFD—350ci V-8 4bbl 175hp—Turbo-Hydramatic
TDD—350ci V-8 4bbl 175hp—manual trans
TDH—350ci V-8 4bbl 175hp—manual trans
TDJ—350ci V-8 4bbl 175hp—Turbo-Hydramatic
TDK—350ci V-8 4bbl 175hp—Turbo-Hydramatic
TAX—350ci V-8 4bbl 175hp—manual trans
VJZ—350ci V-8 4bbl 175hp—manual trans
TKA—402ci V-8 4bbl 210hp—manual trans
TKB—402ci V-8 4bbl 210hp—Turbo-Hydramatic
TKM—402ci V-8 4bbl 210hp—manual trans
TLM—402ci V-8 4bbl 210hp—Turbo-Hydramatic

C-20 Series

TPL—250ci I-6 1bbl 110hp—manual trans
TPY—250ci I-6 1bbl 110hp—Turbo-Hydramatic
TPT—250ci I-6 1bbl 110hp—manual trans
TPW—250ci I-6 1bbl 110hp—manual trans
TPX—250ci I-6 1bbl 110hp—manual trans
TPS—250ci I-6 1bbl 110hp—Turbo-Hydramatic
TPR—250ci I-6 1bbl 110hp—manual trans
THJ—250ci I-6 1bbl 110hp—manual trans
TJA—250ci I-6 1bbl 110hp—manual trans
TJB—250ci I-6 1bbl 110hp—manual trans
TJD—250ci I-6 1bbl 110hp—Turbo-Hydramatic
TJL—250ci I-6 1bbl 110hp—manual trans
TLY—250ci I-6 1bbl 110hp—manual trans
TKR—292ci I-6 1bbl 125hp—Turbo-Hydramatic
TKL—292ci I-6 1bbl 125hp—manual trans
TKP—292ci I-6 1bbl 125hp—manual trans
TDA—307ci V-8 2bbl 135hp—manual trans
TDL—307ci V-8 2bbl 135hp—Turbo-Hydramatic
TRA—307ci V-8 2bbl 135hp—manual trans
TAH—307ci V-8 2bbl 135hp—Turbo-Hydramatic
TJP—307ci V-8 2bbl 135hp—Turbo-Hydramatic
TJR—307ci V-8 2bbl 135hp—manual trans
TSP—307ci V-8 2bbl 135hp—Turbo-Hydramatic
TSS—307ci V-8 2bbl 135hp—manual trans
TBL—350ci V-8 4bbl 175hp—manual trans
TFD—350ci V-8 4bbl 175hp—Turbo-Hydramatic
TDD—350ci V-8 4bbl 175hp—manual trans
TDH—350ci V-8 4bbl 175hp—manual trans
TDJ—350ci V-8 4bbl 175hp—Turbo-Hydramatic
TDK—350ci V-8 4bbl 175hp—Turbo-Hydramatic
TRH—350ci V-8 4bbl 175hp—manual trans
TRJ—350ci V-8 4bbl 175hp—Turbo-Hydramatic
TAX—350ci V-8 4bbl 175hp—manual trans
TAY—350ci V-8 4bbl 175hp—manual trans
VJZ—350ci V-8 4bbl 175hp—manual trans
TKA—402ci V-8 4bbl 210hp—manual trans
TKB—402ci V-8 4bbl 210hp—Turbo-Hydramatic
TKW—402ci V-8 4bbl 210hp—manual trans
TKX—402ci V-8 4bbl 210hp—Turbo-Hydramatic
TKM—402ci V-8 4bbl 210hp—manual trans
TLM—402ci V-8 4bbl 210hp—Turbo-Hydramatic

C-30 Series

TPY—250ci I-6 1bbl 110hp—Turbo-Hydramatic
TPW—250ci I-6 1bbl 110hp—manual trans
TPX—250ci I-6 1bbl 110hp—manual trans
THJ—250ci I-6 1bbl 110hp—manual trans
TJB—250ci I-6 1bbl 110hp—manual trans
TJD—250ci I-6 1bbl 110hp—Turbo-Hydramatic
TJL—250ci I-6 1bbl 110hp—manual trans
TLY—250ci I-6 1bbl 110hp—manual trans
TKS—292ci I-6 1bbl 125hp—Turbo-Hydramatic
TKL—292ci I-6 1bbl 125hp—manual trans
TKP—292ci I-6 1bbl 125hp—manual trans
TDH—350ci V-8 4bbl 175hp—manual trans
TDK—350ci V-8 4bbl 175hp—Turbo-Hydramatic
TRH—350ci V-8 4bbl 175hp—manual trans

TRJ—350ci V-8 4bbl 175hp—Turbo-Hydramatic
TAY—350ci V-8 4bbl 175hp—manual trans
TKW—402ci V-8 4bbl 210hp—manual trans
TKX—402ci V-8 4bbl 210hp—Turbo-Hydramatic

K-10 Series

TPL—250ci I-6 1bbl 110hp—manual trans
TPY—250ci I-6 1bbl 110hp—Turbo-Hydramatic
TPT—250ci I-6 1bbl 110hp—manual trans
TPW—250ci I-6 1bbl 110hp—manual trans
TPS—250ci I-6 1bbl 110hp—Turbo-Hydramatic
TPR—250ci I-6 1bbl 110hp—manual trans
TLY—250ci I-6 1bbl 110hp—manual trans
TDB—307ci V-8 2bbl 135hp—manual trans
TDP—307ci V-8 2bbl 135hp—Turbo-Hydramatic
TAD—307ci V-8 2bbl 135hp—manual trans
TAJ—307ci V-8 2bbl 135hp—Turbo-Hydramatic
TSR—307ci V-8 2bbl 135hp—manual trans
TFH—350ci V-8 4bbl 175hp—manual trans
TFJ—350ci V-8 4bbl 175hp—Turbo-Hydramatic
TDG—350ci V-8 4bbl 175hp—manual trans
TDR—350ci V-8 4bbl 175hp—Turbo-Hydramatic
TFH—350ci V-8 4bbl 175hp—manual trans
TFJ—350ci V-8 4bbl 175hp—Turbo-Hydramatic
TDG—350ci V-8 4bbl 175hp—manual trans
TDR—350ci V-8 4bbl 175hp—Turbo-Hydramatic

K-20 Series

TPL—250ci I-6 1bbl 110hp—manual trans
TPY—250ci I-6 1bbl 110hp—Turbo-Hydramatic
TPT—250ci I-6 1bbl 110hp—manual trans
TPW—250ci I-6 1bbl 110hp—manual trans
TPS—250ci I-6 1bbl 110hp—Turbo-Hydramatic
TPR—250ci I-6 1bbl 110hp—manual trans
TJA—250ci I-6 1bbl 110hp—manual trans
TJB—250ci I-6 1bbl 110hp—manual trans
TJD—250ci I-6 1bbl 110hp—Turbo-Hydramatic
TLY—250ci I-6 1bbl 110hp—manual trans
TKR—292ci I-6 1bbl 125hp—Turbo-Hydramatic
TKL—292ci I-6 1bbl 125hp—manual trans
TDB—307ci V-8 2bbl 135hp—manual trans
TDP—307ci V-8 2bbl 135hp—Turbo-Hydramatic
TAD—307ci V-8 2bbl 135hp—manual trans
TAJ—307ci V-8 2bbl 135hp—Turbo-Hydramatic
TJS—307ci V-8 2bbl 135hp—Turbo-Hydramatic
TJT—307ci V-8 2bbl 135hp—manual trans
TSJ—307ci V-8 2bbl 135hp—Turbo-Hydramatic
TSR—307ci V-8 2bbl 135hp—manual trans
TFH—350ci V-8 4bbl 175hp—manual trans
TFJ—350ci V-8 4bbl 175hp—Turbo-Hydramatic
TDG—350ci V-8 4bbl 175hp—manual trans
TDR—350ci V-8 4bbl 175hp—Turbo-Hydramatic
TRK—350ci V-8 4bbl 175hp—manual trans
TRL—350ci V-8 4bbl 175hp—Turbo-Hydramatic

Transmission Codes

Code	Type	Plant
B	TH	Cleveland
C	Powerglide	Cleveland
H	3 speed	Muncie*
J	TH	GM/CAN
M	3 speed	Muncie
N	4 speed	Muncie
P	4 speed	Muncie
R	4 speed	Saginaw
S	3 speed	Muncie
Y	TH	Toledo

*El Camino

Axle Identification

Code	Ratio

El Camino

CH, GC, GD, GH, KD,	
GF, GN	2.73:1
	3.08:1
CF, CW, RU, RV	3.31:1
GG, GI,	3.36:1

C-10 Series

RHA, RHB, RHS, RHT,	
RRR, RRS, RRT, RRW	3.07:1
RAA, RHC, RHW, RHX	3.73:1
RHD, RHG, RHY, RHZ	4.11:1

C-20 Series

RKH, RKJ, RKP, RKR,	
RKD, RKG	3.54:1
RAH, RAJ, RJA, RJB,	
RKK, RKL, RKS, RKT,	
RJB, RJS, RKB, RKC	4.10:1
RKW, RKX, RRA, RRB,	
RRC, RRG,	4.56:1
RGA, RHP, RJC, RJR,	
RJT,	4.57:1

C-30 Series

RAK, RAL, RAP, RJK	4.10:1
RAC, RAD, RAG, RHR	4.57:1
RAR, RGW, RJL, RRL	5.14:1
RRY	5.43:1
RAS, RAT	6.17:1

K-10 Series

RAY, RAZ	3.07:1
RAW, RAX	3.73:1

K-20 Series

RJH, RJJ	4.10:1
RKY, RKZ	4.11:1
RJD, RJG, RJW,	4.57:1

Exterior Color Codes

C-10, C-20, C-30 Series

Midnight Black	500
Willow Green	504
Spruce Green	506
Hawaiian Blue	510
Spanish Gold	511
Crimson Red	514
Tangier Orange	516
Meadow Green	518
Wheatland Yellow	519
Frost White	521
Classic Bronze	522
Mariner Blue	523
Grapefruit Yellow	525

El Camino

Antique White	11
Pewter Silver	14
Ascot Blue	24
Mulsanne Blue	26
Spring Green	36
Gulf Green	43
Sequoia Green	48
Covert Tan	50
Placer Gold	53
Cream Yellow	56
Golden Brown	57
Mohave Gold	63
Orange Flame	65
Midnight Bronze	68
Cranberry Red	75

Regular Production Options

Vega

14150	Panel Express	$2,080

Luv

82	Mini pickup	2,196

El Camino

13380	El Camino 6cyl	2,790
13480	El Camino V-8	2,880
13680	Custom El Camino V-8	2,960

C-10 Series

CS10514	Blazer	2,700
CS10703	Chassis and cab	2,530
CS10704	Stepside pickup	2,680
CS10734	Fleetside pickup	2,680
CS10904	Stepside pickup	2,715
CS10906	Suburban w/rear doors	3,610

CS10934 Fleetside pickup 2,715
Add $575 for 115" & 127" WB 4WD
models

C-20 Series

CS20903 Chassis and cab 2,760
CS20904 Stepside pickup 2,915
CS20906 Suburban w/rear
    doors 3,767
CS20916 Suburban w/rear
    gate 3,680
CE21024 Longhorn pickup 3,088
Add $660 for 127" WB 4WD models

C-30 Series

CE31003 Chassis & cab 2,962
CS31004 Stepside pickup 3,105

K-5 Blazer

KS10514 Utility pickup 3,258

El Camino

Option Number

AK1 Deluxe seatbelts & front
    shoulder harness
AU3 Electric door locks
A01 Tinted glass (all windows)
A02 Tinted glass (windshield)
A31 Electric windows
A39 Custom deluxe seatbelts
A41 4-way electric control front
    seat
A46 4-way electric control front
    bucket seat
A51 Strato-Bucket seats
A85 Deluxe shoulder harness
B37 Floor mats
B84 Bodyside molding equipment
B85 Upper bodyside molding
B90 Door & window moldings
B93 Door edge guards
C08 Exterior soft trim roof cover
C50 Rear window defroster
C60 Deluxe A/C
D33 Remote control outside LH
    mirror
D34 Visor vanity mirror
D55 Console
D88 Sport stripe
F40 Special front & rear suspension
G41 Special performance front &
    rear suspension
G67 Rear shock absorber level
    control
G80 Positraction rear axle
JL2 Front disc brakes

J50 Vacuum power brake
    equipment
K30 Speed control, cruise master
K85 63 amp AC generator
LS3 400ci (402) V-8 engine
LS5 454ci Hi-Performance V-8
    engine
L48 350ci V-8 engine
L65 350ci V-8 engine
MC1 HD 3 speed transmission
M11 Floor shift transmission
M20 Four speed wide range
    transmission
M22 Four speed transmission heavy
    duty
M35 Powerglide transmission
M38 300 Deluxe 3 speed automatic
    transmission
M40 Turbo-Hydramatic
    transmission
NK2 Deluxe steering wheel
NK4 Deluxe steering wheel
N33 Tilt-type steering wheel
N40 Power steering
PA3 Special wheel trim cover
P01 Wheel trim cover
P02 Deluxe wheel trim cover
T58 Rear wheel opening skirt
T60 Heavy duty battery
UM1 Push-button AM radio & tape
    player
UM2 Push-button AM/FM stereo
    radio & tape player
U14 Instrument panel gauges
U35 Electric clock
U63 Radio—push button control
U69 Radio AM/FM push button
    control
U76 Windshield antenna
U79 AM/FM stereo radio
V01 HD radiator
V30 Front & rear bumper guards
YD1 Axle for trailering
YF3 Heavy Chevy package
ZJ7 Special wheel, hubcap & trim
    ring
ZJ9 Auxiliary lighting group
ZL2 Special ducted hood air system
ZQ9 Performance ratio rear axle
Z15 SS equipment

C10, C20, C30 Series

A07 Glass tinted (10 windows)
A08 Tinted RH body side glass
    (4 windows)
AO9 Laminated glass
A10 Full view rear window
A11 Soft ray tinted glass

A12	Rear window glass	DF2	Mirrors, exterior, stainless steel, camper style
A18	Swing-out rear door glass		
A24	Cab corner windows	DG4	West coast mirror, Jr. stainless steel
A28	Sliding rear window		
A34	Bostrom driver's seat	DG5	West coast mirror, Sr, driver & passenger side (C-30 only)
A35	Bostrom passenger seat		
A50	Bucket seats	DG8	Outside rearview mirror RH (fixed arm)
A52	Custom bench seat		
A54	Full width front seat	DG9	Outside rearview mirror, RH (swinging arm)
A55	Level ride seat		
A56	Bostrom Levelair Bostrom seat	DH3	Outside rear view mirror LH,RH (swinging arm)
A57	Auxiliary seat		
A59	Supplementary seat	E23	HD cab lifting torsion bar
A61	Auxiliary seat, stationary	E28	Assist handles
A62	Front seat belt	E56	Platform and stake rack
A63	Less rear seat belt	E57	Platform body
A78	Center seat	E80	Pickup box mounting brackets
A80	Center and rear seat	E81	Floor board
A85	Shoulder harness	E85	Body side door equipment
A94	Door safety lock	FO3	Heavy duty frame
A97	Spare wheel lock	F19	Special body cross sill mounting support
A99	Instrument panel compartment		
AA2	Tinted windshield glass	F25	Frame rails, full depth, heat treated
AM2	Heavy duty seat		
AM3	Front seat center belt	F43	Front axle, 9000#
AS3	Rear seat	F44	Front axle, 11000#
AS5	Shoulder harness, center and rear seats	F47	Front axle, 5000#
		F48	Front axle, 7000#
AU2	Cargo door lock unit	F49	Heavy duty four wheel drive front axle (K-20)
B30	Floor and toe panel carpet		
B55	Full foam seat cushion	F51	Front and rear HD shock absorbers
B59	Padded seat back frame		
B85	Body side moulding, upper	F54	Front axle, 12000#
B93	Door edge guards	F60	Heavy duty front springs
B98	Side trim molding	F76	Free wheeling front hub
BE2	Padded hinge pillar	G60	Auxiliary rear springs
BX1	Body side molding, belt	G64	Rear spring equipment, 34500#
BX2	Body side molding, wide lower	G68	Rear shock absorbers
C07	Auxiliary top	G70	Rear suspension—leaf spring
C20	Single speed windshield wiper	G80	Rear axle, Positraction
C41	Heater-economy	G86	No-Spin rear axle
C42	Heater Deluxe	G87	Rear axle, power lock
A48	Heater delete	G94	Rear axle, 3.31:1
C60	Air conditioning, all weather	H01	Rear axle, 3.07:1
C69	Air conditioning, roof mounted	H04	Rear axle, 4.11:1
C70	Air conditioning	H05	Rear axle, 3.73:1
D14	Arm rest front door	H20	Rear axle, 4.57:1
D20	Sunshade, windshield, RH	H22	Rear axle, 6.17:1
D23	Sunshade, padded LH	J91	Trailer brake equipment
D29	West coast mirror—Jr	JA1	Rear axle, 4.10:1 ratio (C-20)
D30	West coast mirror—Sr	JA4	Dual brake system
D32	Rear view mirror	K05	Engine block heater
D36	Non-glare inside mirror	K12	Oil filter, 2qt capacity
D48	Front cross view mirror	K28	Fuel filter equipment
D89	Body paint stripe	K29	Vacuum connector
DF1	Mirrors, exterior, painted, camper style	K31	Throttle control, manual
		K37	Engine governor

K43	Air cleaner, dry inside type	U01	Roof marker and identification lamps
K47	Air cleaner equipment		
K48	Oil bath air cleaner	U06	Air horn
K56	Air compressor equipment	U08	Dual horns
K67	Heavy duty starter motor	U16	Tachometer
K76	61 amp Delcotron generator	U60	AM radio—manual
K77	55 amp Delcotron generator	U63	AM radio—pushbutton
K79	42 amp Delcotron generator	U85	Trailer light cable, 6 wire
K81	62 amp Delcotron generator	U86	Trailer jumper cable
K84	47 amp Delcotron generator	U87	Trailer light cable, 7 wire
L05	130 amp Delcotron generator	U92	HD wiring
L22	Engine 250ci L-6	U98	Junction box and wiring
L25	Engine 292ci L-6	UY1	Harness, camper body wiring
L47	Engine 400 (402)ci V-8	V01	HD radiator
M01	Heavy duty clutch	V04	Radiator shutters
M13	Transmission, three-speed HD	V05	HD cooling
M16	Transmission, three-speed HD	V35	Wraparound front bumper
M20	Transmission, four-speed	V37	Chrome front/rear bumpers
M24	Transmission, four-speed HD	V38	Rear bumper—painted
M28	Transmission, four-speed HD close ratio	V43	Rear step bumper, painted
		V46	Front bumper, chrome
M35	Powerglide transmission	V62	Hydraulic jack
M49	Transmission, Turbo-Hydramatic	V66	Provisions for front end drive power take-off
NA4	Fuel tank skid plate	V75	Hazard and marker lights
NO2	Fuel tank—30 gallon	V76	Front tow hooks
N03	Fuel tank, dual, 37 gallon	X56	Front bumper construction
N40	Power steering	X58	Cab insulation
PO1	Wheel trim cover	X59	Cab HD insulation
P03	Hub caps	Z53	Gauges
P10	Spare wheel carrier—under frame mounting	Z58	Auxiliary top
		Z69	Motorhome chassis conversion
P13	Spare wheel carrier—side mounted	Z70	³/₄ ton special 7800# GVW
		Z81	Camper special
T60	Heavy duty battery	Z84	Custom Sport Truck
TP2	Auxiliary battery, camper equipped trucks	ZJ8	Dual rear wheel conversion

Facts

The overall styling of the 1972s was identical to that of the 1971s, however, there were with minor trim changes.

New for the interior was the Highlander Package. It consisted of plaid upholstery in four colors. The Highlander was available with the Custom Deluxe Package on pickups and Suburbans and with the CST Package on the Blazer.

The Cheyenne Super Package included hound's-tooth upholstery trim. A total 40,636 were installed.

The CST Package was installed in 142,645 trucks.

Engine availability was unchanged. Ratings were lower, however, as because output was given as net, not gross, horsepower.

The 350ci V-8 was the most popular engine, with 467,968 sold. The use of automatic transmission was gaining as well, as 413,834 were sold. Total Chevrolet truck sales for the model year amounted to 828,961 units.

This year saw the fewest visual changes on the El Camino: new front turn signal turn lamps, and a three-tiered grille. All other visual SS features were carried over from 1971: SS emblems on the blacked-out grille, the front fenders, and the rear bumper; a domed hood with locking pins; and F60x15 raised-white-letter tires on 15x7 gray Sport wheels.

Power disc brakes were standard, but the F41 suspension was standard only with the 454ci or 402ci engines. It could be purchased at additional cost with the other V-8s.

The SS equipment option was now available with any V-8, including the 130hp 307ci small-block, the two 350ci small-blocks rated at 165hp and 175hp, and the LS3 402ci big-block rated at 240hp. The LS5 454ci V-8 was down to 270hp. Part of the One reason for lower horsepower ratings was that the more realistic SAE net rating system was used. The other reason was that the engines were retuned to meet harsher emission standards.

The 1972 and later El Caminos got a new serial numbering system that included a letter code for the engine installed.

The Vega Panel Express continued with minor trim changes.

A new addition was the Light Utility Vehicle (LUV). It was built by Isuzu Motors, of which GM owned a stake in. The LUV had a 102.4in wheelbase and a 6ft bed, and was powered by a 110.8ci inline four-cylinder engine putting out 75hp. The suspension was independent at the front, and torsion bars and leaf springs at the rear. The axle ratio was 4.56:1, the only tansmission was a four-speed manual and, power drum brakes were standard equipment.